INSTRUCTOR'S MANUAL AND KEY FOR

BASIC MATHEMATICS

SKILLS AND APPLICATIONS

MERLE WOOD
Specialist, Business Education
Oakland Public Schools
Oakland, California

THOMAS HENDRICKS
Teacher
New York City Public Schools
New York, New York

SUSAN MULLER
Sponsoring Editor, Business Mathematics
Gregg Division
McGraw-Hill Book Company
New York, New York

GREGG DIVISION
McGRAW-HILL BOOK COMPANY

New York Atlanta Dallas St. Louis San Francisco
Auckland Bogotá Guatemala Hamburg Johannesburg Lisbon
London Madrid Mexico Montreal New Delhi Panama Paris
San Juan São Paulo Singapore Sydney Tokyo Toronto

CONTENTS

QUESTIONNAIRE iii
THE BASIC MATHEMATICS PROGRAM 1
COMPONENTS OF THE PROGRAM 1
- Skills Test 1
- Textbook 2
- *Learning Guide* 2
- Instructor's Manual and Key 2

CONTENT COVERAGE 2
- Mathematics Competencies 2
- Business and Consumer Topics 3

INDIVIDUALIZED OR GROUP INSTRUCTION 3
MANAGEMENT SYSTEM 6
- Monitoring Progress 6
- Tests Results Become Assignments 6
- Instructional Process 6
- What Testing Tells You So Far 7
- Beginning Instruction 8
- Goal Practice and Goal Reviews 8
- Skill Analysis: Error Pattern Remediation 8
- Self-Checking 8
- Auditing, Business, and Consumer Corners 8
- The *Learning Guide:* Review, Evaluation, Enrichment 8
- Completing One Goal/Beginning the Next 8

WORD PROBLEMS 8
ERROR PATTERNS 10
SPEED DRILLS 13
ANSWERS TO SPEED DRILLS 18
TEXTBOOK ANSWER KEY 24
ANSWER KEY TO LEARNING GUIDE
- Skill Builders 54
- Projects 60
- Extra Practices 72
- Quizzes 82

INSTRUCTOR'S ANSWER KEY TO PRECHECKS, GOAL EVALUATIONS, AND ALTERNATE GOAL EVALUATIONS 86
GOAL EVALUATIONS 97
ALTERNATE GOAL EVALUATIONS 143
TRACK RECORD 191

Instructor's Manual and Key for *Basic Mathematics: Skills and Applications.*

1 2 3 4 5 6 7 8 9 0 EDWEDW 8 9 0 9 8 7 6 5 4 3

ISBN 0-07-071603-X

Questionnaire for Basic Mathematics: Skills and Applications

Dear Instructor:

Would you like to help us improve the quality and usefulness of your textbooks? We hope so. We are very interested in what you think of this new program—*Basic Mathematics: Skills and Applications.* Your comments and suggestions will help us plan new or revised programs to better meet your needs. If you would take a few minutes to jot down your reactions after you have used this program, they would be most welcome. Please return the questionnaire to the **Editor-in-Chief, Business Mathematics, Gregg Division, 29th Floor, McGraw-Hill Book Company, 1221 Avenue of the Americas, New York, New York 10020.**

Basic Mathematics: Skills and Applications (BMSA) has four components that can be used together or separately: the textbook, the *Learning Guide,* the *Skills Test,* and the Instructor's Manual and Key.

Textbook (grade level of students in this course ________)

1. Content coverage: Please rate the coverage of these aspects of the textbook.

	Comprehensive	Above average	Sufficient	Insufficient
a. Math skills covered				
b. Business and consumer topics (Auditing, Business, and Consumer Corners)				
c. Solving word problems				

2. Pedagogy: Please rate the effectiveness of these techniques and features of the textbook.

	Excellent	Good	Average	Below average
a. Short units of instruction				
b. Step-by-step explanations				
c. Word Problem Strategies				
d. Skill Analysis Problems and Appendix				

3. Practice problems: rate the quantity and difficulty of the practice problems.

	Quantity			Difficulty Level		
	Plenty	Enough	Need more	Right level	Too easy	Too hard
a. Goal Practices/Reviews						
b. Auditing, Business, and Consumer Corners						

Learning Guide

4. Which features of the *Learning Guide* did you use? Did you find them effective?

	Used	Effective
a. Prechecks		
b. Quizzes		
c. Skill Builders		
d. Projects		
e. Extra Practices		
f. Track Record		

5. Please rate the Extra Practice exercises found at the end of the *Learning Guide.*

a. Quantity: Plenty ____ Enough ____ Need more ____

b. Difficulty: Right level ____ Too easy ____ Too hard ____

Skills Test

The *Skills Test* was developed by Gregg/McGraw-Hill in consultation with CTB/McGraw-Hill. It is available for sale in sets of 25.

6. How was the *Skills Test* used?

As a pretest ____ As a posttest ____ As both ____ Not at all ____

7. How much time did it take to give the test once? Hours ____ Minutes ____

8. Into how many sessions did you divide the administration of the test? ____

9. Was the test helpful

a. In planning instruction? Yes ____ No ____ **b.** In evaluating progress? Yes ____ No ____

10. Was the Class Record useful? Yes ____ No ____
(See the Teacher's Manual to the *Skills Test*)

Course Management and the Instructor's Manual and Key

11. Which features of the IMK did you use? Did you find them effective?

Feature	Used	Effective	Feature	Used	Effective
a. Speed Drills	______	______	c. Alternate Goal Evaluations	______	______
b. Goal Evaluations	______	______	d. Track Record	______	______

12. Did you make the answer key to practice problems available to students?

Yes ______ No ______

13. What percentage of instruction was accomplished

a. individually ______%

b. in small groups ______%

c. with the whole class ______%

\+ ______(100%)

14. How was the Track Record used?

a. Teacher recorded assignments and scores ______

b. Students recorded assignments and scores ______

c. Both a and b ______

d. Not used ______

15. What is your overall rating of this program?

Excellent ______ Good ______ Average ______ Poor ______

a. What features do you like most about this program?

______

______

b. What features do you like least about this program?

______

______

16. Do you have any suggestions for improving this program?

Add: ______

______

Delete: ______

______

Other: ______

______

17. a. Do you use or plan to use a microcomputer in this course?

Yes, I use now ______

No, I do not use now ______

No, but I will use in the next 4 years ______

Other ______

b. If microcomputer equipment is or will be available, what types of equipment do you use or plan to use? ______

c. For what purposes do you or will you use microcomputer in this course?

______ Remediation in basic arithmetic skills

______ course management

______ testing

______ Computer-assisted instructions in learning business mathematics

______ business applications

______ drill and practice

______ Other ______

18. Thanks for your help. If you might be interested in acting as a reviewer for us in the future, please fill in the information below.

Name ______ Institution ______

Address ______ Phone ______

THE BASIC MATHEMATICS PROGRAM

The goal of the *Basic Mathematics* program is to build basic mathematical skills essential for success in the work world and in everyday life. The program was designed to provide the following:

Crucial Content
- Instruction and practice on basic mathematical skills needed in entry-level office jobs
- Instruction and practice in important business and consumer topics

Sound Teaching Techniques
- Diagnosis of student needs
- Small instructional steps that explain basic arithmetic concepts
- Reteaching of skills learned incorrectly
- Review of partially mastered skills
- Skill-based instruction

Efficient Management
- Format directing students through the right material at their own individual pace
- Recordkeeping that is part of the instructional process

Flexible Use—Core or Supplementary Textbook
The *Basic Mathematics* program may be used as either a core or supplementary textbook. In business mathematics, consumer mathematics, or remedial mathematics classes, this program can serve as the core of the curriculum. Where students are having difficulty in basic office education or distributive education courses because of math deficiencies, this program can provide on-the-spot remediation.

• COMPONENTS OF THE PROGRAM

The *Basic Mathematics* program consists of four separate items: (1) the *Skills Test*, which may be used as both a pretest and a posttest, (2) the textbook, (3) the *Learning Guide*, and (4) the *Instructor's Manual and Key*.

Skills Test

The *Skills Test* is designed to be used both as a pretest and a posttest for the whole class. As the first step in the *Basic Mathematics* program, the *Skills Test* tests

COMPONENTS OF THE BASIC MATHEMATICS PROGRAM

SKILLS TEST	LEARNING GUIDE	TEXTBOOK	INSTRUCTOR'S MANUAL AND KEY
4 test items for each of 25 skills (100 items total) *Includes:* 25 copies of *Skills Test* Scoring template Teacher's Manual (including class record)	Goal Prechecks Skill Builders Section Quizzes Enrichment Projects Extra Practice problems *Includes:* Student copy of *Track Record* (recordkeeping and assignment sheet)	Instruction Practice Remediation Review Business and consumer applications Problem Solving	Teaching and remediation suggestions Two sets of unit posttests (Goal Evaluations and Alternate Goal Evaluations) Answers to all tests Answers to all practice problems (may be given to students for self-checking)

students' knowledge of the 25 mathematics skill areas covered by the program. A scoring template, class record, and teacher's manual are included in the package of test booklets. Used as a pretest, the *Skills Test* will identify skill deficiencies and prescribe appropriate work in the textbook and *Learning Guide*.

At the end of the course the *Skills Test* can be given again as a posttest to evaluate student progress and provide documentation of the class's accomplishments.

Textbook

The 25 chapters (Goals) in the text cover the essential mathematics skills and relevant business and consumer applications. Each Goal is divided into Sections and lessons, providing manageable units of instruction. Each lesson is followed by Goal Practice and Goal Review exercises. These exercises consist of 6 to 20 problems covering only the material taught in that lesson. If the student completes the Goal Practice perfectly, he or she is instructed to skip to the next lesson. If not, the Goal Review is assigned.

Auditing, Business, and Consumer Corners

Each Section of a Goal is followed by Auditing, Business, and Consumer Corners. These provide applications of the material covered in the Goal in an appropriate business or consumer setting. In the Auditing Corner students are presented with problems that have been solved and are asked to check and correct any errors. The Business and Consumer Corners provide a brief introduction to a topic and then a series of problems about the topic. The Business and Consumer Corners utilize business forms where appropriate.

Skill Analysis: Error Pattern Remediation

The textbook also contains several Skill Analyses. These lessons provide error pattern remediation. Problems with multiple-choice answers are used to catch improper problem-solving methods. Each incorrect answer is then analyzed to show how the incorrect answer was reached and to teach the correct methods.

Word Problem Strategies

Several of these lessons, designed for group instruction, appear throughout the text and are reinforced in Goal 19 (Solving Word Problems).

Learning Guide

The *Learning Guide* contains the Goal Prechecks, Skill Builders, Section Quizzes, Projects, Extra Practice exercises for each lesson, and the student Track Record.

Precheck

The Precheck is a pretest for each mathematics competency. After taking the *Skills Test* as a pretest, students take the Precheck for those Goals on which they scored below the mastery score. The Precheck provides a more comprehensive evaluation of a student's knowledge of a competency and permits you to pinpoint specific areas of weakness. These are the lessons of the Goal that the student needs to cover. Students may skip those Goals or lessons in the textbook for which their Precheck scores indicate mastery of the content.

Skill Builders

The Skill Builders are supplementary exercises that review and reinforce the math skills covered in the Goal. Often these are games or puzzles designed to interest the student.

Section Quizzes

The Section Quizzes review the material covered in each Section and determine if a student is ready to move on to the next Section or to take the Goal Evaluation.

Projects

These enrichment projects teach higher-level skills. They can be used for group instruction or to provide enrichment for students who are more advanced than the rest of the class.

Track Record

The Track Record is used to record pretest and Precheck results and assignments. It also tracks progress in mastering the goals of the program. The Track Record is the student's guide to the use of the program.

Instructor's Manual and Key

The *Instructor's Manual and Key* includes teaching and remediation suggestions, answers for tests and practice problems, and two sets of Goal Evaluations (unit posttests). The answer key for the practice problems is in a separate section and may be removed or duplicated and distributed to students so that they can check their own work. Answers for the exercises in the *Learning Guide* and for all tests are also included.

• CONTENT COVERAGE

Extensive research has been done to identify the mathematics competencies needed in entry-level business and office careers. Job tasks were analyzed. Various state and local competency listings and common business and office education as well as distributive education textbooks were carefully examined.

Mathematics Competencies

From this research 25 mathematics skills were selected as important to students preparing for entry-level jobs and for further business education. Nineteen competencies were ranked as essential to most business and office careers; another six competencies were deemed of secondary importance to business and consumer concerns. Listed below are the 25 competencies that form the basis of the program.

Whole and Decimal Numbers

1.* Express a number in numeric form and in word form.
2.* Round off numbers and money amounts to the nearest specified place.
3.* Add four five-digit whole numbers.
4.* Subtract a five-digit whole number from a five-digit whole number.
5.* Multiply a four-digit whole number by a three-digit whole number.
6.* Divide a five-digit whole number by a three-digit whole number.
7.* Position the decimal point in an answer to an addition, subtraction, multiplication, or division problem.
8.* Express a number as a common fraction, a decimal fraction, and as a percent.
9.* Rename a common fraction, a decimal fraction, and a percent as another type of fraction.

Measurement

10.* Determine the time elapsed without the use of a calendar or clock.
11.* Identify the approximate size of common U.S. customary units of length, weight, temperature, and capacity.
12.* For a given U.S. customary unit of length, weight, and capacity, select an equivalent U.S. customary unit.
13.* Identify the approximate size of common SI units of length, mass, temperature, and capacity.
14.* For a given SI unit of length, mass, or capacity, select an equivalent SI unit.
15.* With a ruler, measure common physical objects to the nearest $\frac{1}{4}$ inch or millimeter.

Estimation and Word Problems

16.* Estimate the answer to a problem.
17.* Read tables and graphs.
18.* Solve a word problem using a formula for decrease, increase, discount, interest, markup, commission, or percent.
19.* Solve a word problem requiring one step to its solution.

Fractions

20. Express a common fraction in lowest terms.
21. Express a mixed number as a common fraction. (Denominators restricted to 2, 3, 4, 5, 6, 8, 10.)
22. Multiply a five-digit whole number by a common fraction. (Denominators restricted to 2, 3, 4, 5, 8, or 10.)
23. Divide a five-digit whole number by a common fraction. (Denominators restricted to 2, 3, 4, 5, 8, or 10.)
24. Add four common fractions. (Denominators restricted to 2, 3, 4, 5, 8, 10.)
25. Subtract one common fraction from another common fraction. (Denominators restricted to 2, 3, 4, 5, 8, 10.)

*Essential to most business and office careers.

Business and Consumer Topics

Students are exposed to a wide variety of business and consumer applications of the mathematics skills covered in this program. Specific forms and concepts are taught within the textbook Goals in Auditing, Business, and Consumer Corners (see next page). The level is kept simple so that students may comfortably reinforce what they have learned. General and more involved topics are covered in Projects in the *Learning Guide*.

• INDIVIDUALIZED OR GROUP INSTRUCTION

Basic Mathematics is designed to allow you flexibility in individualizing instruction. Individualization can be seen as a scale ranging from large class instruction to personal tutorial. The selection of materials and the pace of instruction are chosen to meet the needs of the group or the individual, but in either case the student is responsible for completing assignments and tests individually.

In large group instruction students often benefit from the security offered by the class structure. Students make few decisions but can interact with each other to enhance learning. The slower or faster students, however, may be lost to boredom or frustration.

In a tutorial setting students receive personally tailored instruction. They may set their own pace and make decisions about their learning needs. A totally individualized approach, however, can isolate students and overwhelm your resources. In teaching *Basic Mathematics* you can strike a balance on this scale. Here are some suggestions to help maximize effectiveness in a semitutorial setting.

1. Minimize the time you spend checking practice problems and tutoring.
   - Make available in class the answer key to Goal Practice and Goal Review problems.
   - Group students in learning teams of two to four. Students can read material and work out examples together. Students should work through practice problems independently but tutor each other when possible.
   - Circulate through the class to spot-check student work and to instruct as needed.

Business and Consumer Topics Taught in the Basic Mathematics Program

In Goal	*Textbook Corners*	*After Goal*	*Learning Guide Projects*
1	Proofreading financial reports Business forms and accounting paper Writing amounts on checks		
2	Rounding off amounts on financial statements Rounding off discounts Rounding off on personal money records		
3	Marketing surveys Sales reports and inventory reports Budgets Expense reports and crossfooting Assets and liabilities Buying a new car		
4	Inventory cards Budgeted versus actual expenses Odometer readings Making change Accounts payable and receivable Check stubs	4	Completing expense reports
5	Recycling scrap Inventory reports Installment buying IRS auditing Purchase orders and invoices Mortgage loans		
6	Checking unit costs Average expenses and average income Unit prices, loss leaders and sales items, the giant ecomony size Checking payroll records Inventory value, unit costs, average costs, merchandise turnover rate Miles per gallon		
7	Double-checking facts Cost control Paycheck deductions Planning and research Bonds and promissory notes Utility bills		
8	Business statistics Planning and small business Hidden costs of consumer products: packaging, processing, and advertising	8	Reconciling a bank statement with checkbook records
9	Translating data for computer entry Analyzing census data Inflation and savings		
10	Verifying a bill Paying for computer time Time zones and military time Organizing work schedules Loan due dates and discount periods Average daily balance		
11	Checking customary abbreviations on business forms Using fertilizer on farmland Choosing an air conditioner		

Business and Consumer Topics Taught in the Basic Mathematics Program (continued)

In Goal	*Textbook Corners*	*After Goal*	*Learning Guide Projects*
12	Changing recipe amounts	12	Becoming computer literate
	Area		
	Perimeter		
13	Checking SI units on business forms		
	Choosing correct SI units		
	Choosing correct SI units		
	Checking SI units on business forms		
	Choosing correct SI units		
	Choosing correct SI units		
	Checking SI units on business forms		
	Choosing correct SI units		
	Choosing correct SI units		
14	Packaging goods for resale		
	Cooperative food stores		
	Menu planning		
15	Checking layouts		
	Scale drawings		
	Map reading		
16	Estimating for sense		
	Estimating for business sense		
	Let the buyer beware		
17	Price schedules for printing and copying	17	Filing an income tax return
	Social security tax tables		
	Life insurance tables		
	Checking graphs for data		
	Customer spending patterns		
	Stock charts		
18	Checking use of the percentage formulas		
	Using key words and formulas to solve business word problems		
	Using key words and formulas to solve consumer word problems		
19	Judging the best solution	19	Making investments
	Seeking more information		
	Getting the facts		
20	The cost of free repairs		
	Public opinion polls		
	Nutritional values of food		
21	Investing in a risky business		
	Ordering fractional quantities		
	Decorating		
22	Checking payroll records, piece-rate plan, hourly rate plan, overtime pay		
	Stock purchases, sharing profit and loss—profits—losses		
	Reduced prices, finding the circumference and area of a circle		
23	Packing goods for shipment		
	Parking lots		
	Yield of gardens		
24	Checking specifications sheets	24	Using different payroll plans
	Time cards		
	Exercise marathon		
25	Food services		
	Fabric inventories		
	Changes in stock prices		

2. Keep students accountable for their work.
   - Make Goal Evaluations and the Skills Test secure: do *not* teach items on these tests or allow students to score these tests.
   - Tell students that these tests are checkpoints at which they can prove to themselves and to you that they have mastered each skill.
3. Keep track of student progress and assignments.
   - Mark Prechecks, Goal Evaluations, and the *Skills Test* yourself.
   - Keep class and/or individual records.
4. Maintain individual and group motivation.
   - Have students keep their own records (students copy scores and assignments onto their Track Records); this device will chart their progress as well.
   - Bring the class together periodically for traditional lessons, class skill games, and discussion of overall progress.
   - You may choose to base part of each student's grade (5 or 10 percent, for instance) on the success of the group in which he or she belongs.

The diversity of students' abilities and their level of responsibility will help you determine the degree of individualization desirable in a class. To keep a comfortable level of control and security for both you and your students, you can gradually change the roles played in the classroom. These three stages of individualization are presented as guidelines for a transition toward personalized instruction in *Basic Mathematics*.

Stage 1—Roles for Traditional Instruction

Instructor	*Students*
• Presents materials	• Participate in discussions
• Answers questions	• Complete assignments
• Assigns work	• Score Practices and Quizzes
• Administers tests	• Keep Track Record
• Scores "secure" tests	

Stage 2—Transitional Role

Instructor	*Students*
• Daily presents selective skills, business topics, word problem strategies, and projects.	• Work at own pace (ahead of or behind presentations) • Tutor peers as directed

Stage 3—Roles on More Personalized Instruction

Instructor	*Students*
• Periodically presents word problem strategies, selected topics, projects, and games for whole class work	• Participate in class activities
	• Work in groups and individually
	• Tutor peers regularly
• Supervises individual and group progress	

At stage 1 you familiarize the class with the procedures of testing, learning, practicing, retesting, and recordkeeping. The class moves ahead as a single group.

At stage 2 you assign students to groups for peer tutoring, and students move at their own pace. You select topics to review for the entire class for part of each period.

At stage 3 you present topics periodically to bring the whole class together. Students routinely work individually in peer tutoring groups. You circulate as a resource to students.

As students become familiar with the structure of the material and the course, more individualization becomes possible. We recommend that you assign deadlines for students to meet in each Goal.

• MANAGEMENT SYSTEM

More and more, educators are being asked to address individual needs and to document individual progress. This program helps instructors to manage these two tasks.

Monitoring Progress

The program management device, the Track Record, is used by you and your students. You record initial *Skills Test* results on a copy of the Track Record (see the duplicatable master on the last two pages of this Manual).

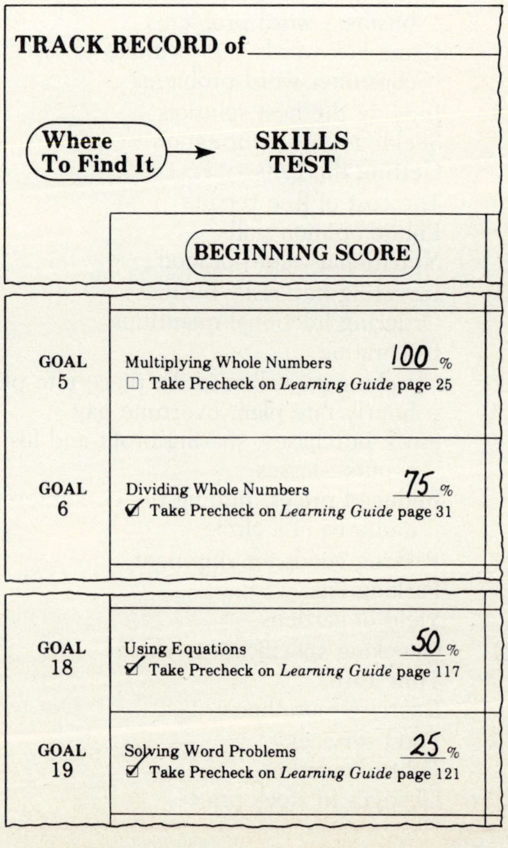

Students record these results on their copy of the Track Record printed on the inside covers of the *Learning Guide*. Other test results are recorded as the students progress. The Track Record then assigns work based on student performance.

Test Results Become Assignments

The results of the *Skills Test* assign students to begin certain units where instruction is needed. A Goal Precheck confirms which Goals to complete. After each Section a Quiz routes students ahead to continue instruction or back to review. At the end of each Goal a Goal Evaluation (posttest) shows progress and directs students to the next Goal. Successful completion of all assigned units leads to the posttest *Skills Test*, which tests overall progress.

The Instructional Process

The pretest *Skills Test* was developed in conjunction with the California Test Bureau (CTB). It is composed of 25 four-item subtests. Each subtest examines a single math skill competency. The teacher records the subtest score on the Track Record indicating:

- Percent correct (0%, 25%, 50%, 75%, or 100%) and
- Need for instruction (√).

A subtest score of less than 75 percent would require a student to take that Goal's Precheck. The students or the instructor can record the pretest results on their Track Records located in the *Learning Guide*. Then the students should be instructed to take the first Goal Precheck you indicated with a check (√) for that Goal.

Students turn to the *Learning Guide* page for their first Goal Precheck and work at their own rates but without assistance.

At the end of each Precheck, there is a place to record the student's score that is a facsimile of his or her Track Record. Indicate the following information on this facsimile and direct the student to copy on his or her Track Record:

- Percent correct in each Section
- Need for instruction

A score below 75 percent on a Section of a Precheck shows a need to study that Section. Mark the textbook Sections to be completed with a check.

What Testing Tells You So Far

Your recordkeeping and students' efforts on the *Skills Test* pretest pay off. You now have a record of group and individual skill needs. The Goal Precheck has told you which Section(s) of the textbook will best help each student achieve competence in a Goal.

TRACK RECORD of *Student's Name*

Where To Find It →	SKILLS TEST	LEARNING GUIDE	TEXTBOOK
	BEGINNING SCORE	PRECHECK	WORK TO DO
GOAL 5	Multiplying Whole Numbers 100 % ☐ Take Precheck on *Learning Guide* page 25	Section 1 ____ % 2 ____ % ☑ NEXT GOAL	☐ Text page 88 ☐ Text page 98
GOAL 6	Dividing Whole Numbers 75 % ☑ Take Precheck on *Learning Guide* page 31	Section 1 100 % 2 60 % ☐ NEXT GOAL	☐ Text page 114 ☑ Text page 133
GOAL 18	Using Equations 50 % ☑ Take Precheck on *Learning Guide* page 117	60 % ☐ NEXT GOAL	☑ Text page 387
GOAL 19	Solving Word Problems 25 % ☑ Take Precheck on *Learning Guide* page 121	40 % ☐ NEXT GOAL	☑ Text page 416

Beginning Instruction

The Track Record directs students to begin work on a specific textbook page. Material may be presented to groups, or students may work individually. Students should be able to read the textbook with little assistance. Diagrams and examples parallel verbal explanations. Short Goal Practices follow small lessons in each Goal. In many instances Goal Practices are introduced with a series of partially solved problems for students to complete. If the textbook is used individually, you should urge each student to reread and rework problems before calling for assistance. (Of course, instructor judgment is essential.)

Goal Practice and Goal Reviews

Typically the Goal Practice consists of examples, number problems, and word problems. Successful completion of the Goal Practice allows the student to skip the Goal Review. The Goal Review typically consists of number problems and word problems. The usual passing criteria recommended are as follows:

- Goal Practice: 100%
- Goal Review: 80%

Students who still have problems after a Goal Review are routed back to you. Students progress through instruction and Goal Practices and Reviews to the end of a section.

Skill Analysis: Error Pattern Remediation

Some students learn a mathematics method incorrectly. In other words, they use the wrong steps to solve a problem. Sometimes these steps give the right answers. Therefore, to remedy such wrong learning requires analysis of the pattern of errors a student makes. This program tries to identify these patterns with carefully chosen multiple-choice items. Each item is designed to "catch" a faulty method. Students are directed to a textbook page according to their responses. On those pages correct methods are presented step by step. Skill Analysis questions appear in selected basic arithmetic skills Goals before Auditing, Business, and Consumer Corners.

Self-Checking

The answer key for all practices may be distributed to each student or stored in the classroom. Since students will be evaluated at the end of each Section and again at the end of each Goal, it is not necessary for you to correct practice assignments. Of course, you may want to check some students' work periodically before Quizzes and tests.

Auditing, Business, and Consumer Corners

Each textbook Section of a Goal culminates in three practical applications of the skills learned. Business and consumer concepts are taught and practiced.

- Auditing Corners present completed problems which students must check and correct.
- Business and Consumer Corners present topics and problems in real-life settings for students to solve. Where appropriate, business forms are used to familiarize students with their use.

The *Learning Guide:* Review, Evaluation, Enrichment

After students finish a Goal Section they are directed to the *Learning Guide*. In this book students prepare for the Goal Evaluation and complete enrichment activities. A Skill Builder, containing a change-of-pace activity, precedes each Quiz serving as a review of skills presented. A Quiz helps students and teachers to decide if students are ready for the next Section or the Goal Evaluation.

Between Goals

Activities that teach problem solving and extensive business topics are located "between" certain Goals. Word Problem Strategies are taught in the textbook. These activities are designed for group instruction and discussion and are reinforced in Goal 19, Solving Word Problems. "Projects" appear in the *Learning Guide* and may be completed individually or serve as group lessons. Group instruction for these higher-level skills may be used to break up long stretches of individualized skill work.

Completing One Goal/Beginning the Next

You have been provided with two duplicatable tests for each Goal. Students take the Goal Evaluation after completing all the Sections of a Goal. The result is recorded on the Track Record. If a student scores less than 75 percent in any part of the test, you and the student plan review. Extra Practices (located in the *Learning Guide*) can come in handy here. When students pass this test, they proceed to the next Goal Precheck identified on their Track Record. The Alternate Goal Evaluations may be used as retests or as secure pretests for each Goal.

• WORD PROBLEMS

Testing throughout the United States has revealed that many students have mastered basic operational skills but are unable to successfully solve word problems. The second National Assessment of Educational Progress in 1978 revealed that the average score on problem solving was only 38 percent correct. While students did reasonably well in one-step problems that could be solved by addition, subtraction, multiplication, or division, "almost every problem that couldn't be solved by a routine application of a single arithmetic operation caused a great deal of difficulty."[1]

[1] Thomas P. Carpenter et al., "Solving Verbal Problems: Results and Implications from National Assessment," *Arithmetic Teacher* (September 1980), p. 8.

In addition to the results of this national testing, the National Council of Teachers of Mathematics recommends that "problem solving be the focus of school mathematics in the 1980s."[2] Why is there such a focus on problem solving? Perhaps because arithmetic facts are not useful unless they are applied to everyday problems. Perhaps also because problem solving is not just the end of mathematics study but also one of the most basic skills requisite in society in almost every pursuit.

Word problems are a type of problem-solving activity found in almost all mathematics textbooks. While life does not offer such neatly packaged problems to solve, word problems are an attempt to simulate real-life situations where math skills are needed. Being able to solve word problems is a connecting bridge between knowledge of basic arithmetic facts and applying those facts to everyday situations.

Teaching word problems isn't easy because what one is trying to do is to teach a student to analyze, to compare, to judge, to question. In other words, one is trying to teach students to think. While this is an admirable goal, it isn't easy, and some instructional copy in a textbook packaged with many problems to solve is not going to accomplish this goal.

What *Basic Mathematics* offers on word problems is as follows:

- A mix of computational and word problems (throughout all Goals in the textbook)
- A framework (the TEST method) to use when solving problems (Goal 19 of the textbook)
- A mix of problems where students are challenged beyond the normal routine of word problems into analyzing whether the problem can be solved, what information is missing, whether there is extraneous information supplied, and other thought-provoking types of questions
- Throughout the textbook, a variety of Word Problem Strategies, which offer students problems with a mix of operations (10 Word Problem Strategies between Goals in the textbook)

The framework offered the student—the TEST method—is not a magic recipe; it is simply a set of procedures designed to give students a grasp of how to solve word problems. This framework is simply a starting point for instructor/student interaction in finding strategies for solving word problems.

Many textbooks offer computational exercises and word problems on a skill-by-skill basis. The disadvantage of this is that students know which operation to use when confronted with a word problem: It's simply the operation they used in solving the number problems. To a certain degree *Basic Mathematics* is forced into this because the structure has a skill-by-skill basis. *Basic Mathematics* deals with this problem by having mixed operations placed throughout the book between the Goals in the Word Problem Strategies and by having a number of Goals that involve a number of different operational skills (word problems, formulas, placing the decimal point, measurement, etc.).

Unfortunately, word problems do not exhaust problem solving but merely constitute one type of problem-solving experience—in fact, a limited one. Word problems don't give us real-world problems where one must think and come up with new strategies for solving problems. However, word problems can function as a bridge between computational skills and more generalized problem solving. But they are only a part of that bridge; the other more substantial part is the instructor, who needs to challenge students to think, to expose them to nonroutine situations, and to show them a variety of ways of handling problem-solving situations.

If word problems are difficult to teach, problem solving is even more difficult because it involves more open-ended activities and allows a variety of solutions. There is no correct way to teach problem solving; by its nature it demands exploration of the nonroutine. Suggestions for doing this are constantly appearing in the literature and being discussed at meetings throughout the country. A number of ways of handling problem solving are discussed below. These suggestions are not new or even original approaches. They are simply a collection of the ideas and recommendations now under discussion. Approach them with an open mind; use whatever will help and work in your classroom.

1. **Open-Ended Topics.** Assign open-ended topics to the class, and have students make up problems based on the topics. The students have to decide what facts are needed and to make up their own problems. Then they decide what operations to use. There is no "right" answer, and the students are free to go in any direction at all. It is a good group activity because other students help each other learn.
   - How much does it cost to go away for a weekend? (How many people are going? Is there more than one way to travel, such as bus, car, train, or plane? What would each cost? What will living costs be? Where would you stay? What would you do?)
   - What would a class dinner cost?
   - How many hours do you have to work to buy a winter coat?
2. **Problems Involving a Graph or Diagram.** Have students make up problems from a graph or diagram. Use this procedure to have them analyze the relationships between the facts.
   - How many more calories are in a cup of cocoa than a cup of cider? An ice cream cone than a brownie? A cup of cider than a brownie?

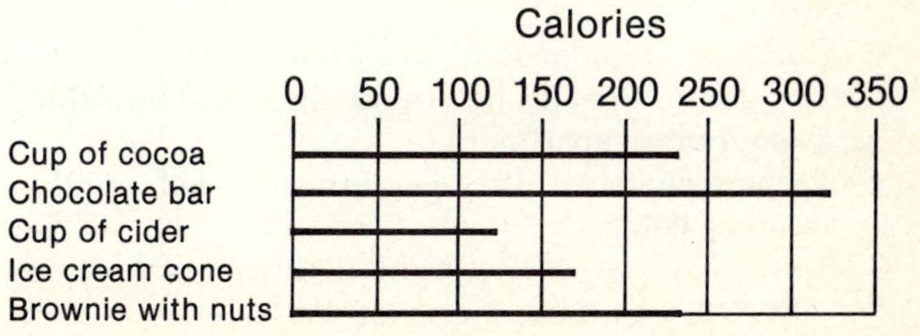

[2] *An Agenda for Action—Recommendations for School Mathematics of the 1980s* (Reston, Va.: National Council of Teachers of Mathematics, 1980), p. 1.

- How many calories are in a cup of cider and a brownie? An ice cream cone and a cup of cider? A chocolate bar and a brownie?

3. **Problems in Which Numbers Are Not Given.** Assign problems in which numbers are not given. Have the students decide what facts are needed and what operations one must perform.
   - How much did Jean spend on slacks and jogging shoes? Possible leading questions might be, What did each cost? Is there a limit on how much she had to spend? Is there more than one price level for each? How many pairs of each did she want?
   - Mildred bought two computer programs at the store where she works. Considering her employee discount, what did she pay for the programs?
4. **Student-Constructed Problems.** Have the students make up problems of their own using a particular operation. This will train them in figuring what kinds of facts are needed and how to "play in" the facts to the operation.
5. **Problems beyond the Scope of the Course.** Have groups work on problems which go beyond the scope of the course. Use them to develop thinking of ways of solving problems.
   - Find two numbers whose square equals another number, such as 25 (3 and 4), 40 (6 and 2), 45 (6 and 3).
   - Find the next number in a sequence. 1, 4, 9, 16, 25, ? (36; they are squares). 1, 2, 3, 5, 8, ? (13; the last two factors are added to the next factor).
6. **File of Interesting Problems.** Make up a file of interesting problems from other textbooks, newspaper ads, news stories, and puzzle books. Save them for students who need extra help in an area or as enrichment material.
7. **Problems Using a Calculator.** Have the class solve problems using calculators. This takes away the possibility of difficulty with the use of arithmetic and concentrates the attention on what operations to use and how the given facts are related to each other.
   - How many seconds are there in a year?
   - How many inches are there from your city to another? To the moon? Around the earth?
   - How many hours since a historical event?
   - How many minutes since the first day of school this term?
   - Measure your hand. How far would 945,342 hands reach?
   - Given a particular cost of gas and a particular car's mileage, how much would it cost to drive 50 yards?
   - If a shower uses 25 gallons of water, how much water do you use in a year taking showers?
8. **Problems from Newspapers.** Use interesting items from the newspapers to provide realistic, exciting problems.
   - Compare attendance figures from major events.
   - Using furniture ads, determine the cost of furnishing a room or an apartment.
   - Prepare averages of temperatures, rainfall, or other weather data.,
   - Determine savings by using discounts or by using different loan rates.
   - Compare the cost of buying the week's groceries from different stores by using their weekly ads.
   - Figure total costs (plus tax, minus discount).
   - Compare the cost of buying on one-, three-, and four-year loans.
   - Compare costs of renting cars of various sizes from different companies (for one day, for a 150-mile trip, for a 300-mile trip, etc.).
   - Compare prices on sales, and determine the percent of savings.
   - Calculate problems.

In using these techniques, it is important not to help the student too directly. This type of activity is meant to stimulate the students' thought processes and to have them learn new ways of attacking problems. If you give the students the answer, they do not learn to think it through for themselves. In general, it is better to use leading questions that make students deduce the answer for themselves.

Bibliography

Arithmetic Teacher (November 1977). Special issue on problem solving. See especially the article by Marilyn N. Suydam and J. Fred Weaver, "Research on Problem Solving: Implications for Elementary School Classrooms."

Carpenter, Thomas P., Mary Kay Corbitt, Henry S. Kepner, Jr., Mary Montgomery Lindquist, and Robert E. Reys. "Solving Verbal Problems: Results and Implications from National Assessment." *Arithmetic Teacher* (September 1980), pp. 8–12.

——. "Results and Implications of the Second Mathematics Assessment: Elementary School." *Arithmetic Teacher* (April 1980), pp. 10–12, 44–47.

Fennell, Francis M. *Elementary Mathematics Diagnosis and Correction Kit.* West Nyack, N.Y.: Center for Applied Research in Education, 1981.

Muller, Adelyn C., and Ray Kurtz. "Students Like Personalized Word Problems." *Arithmetic Teacher* (May 1981), pp. 13–14.

Polya, G. *How to Solve It.* Princeton, N.J.: Princeton University Press, 1973.

• ERROR PATTERNS

It's easy to tell when students make a mistake in mathematics; the answer is wrong. Correcting the mistake, however, is not that easy. Researchers in arithmetic remediation have discovered that one of the best ways to help students avoid mistakes is by discovering why the mistake was made. Many factors have been identified, but one of the most common is that the student is solving the problem in a faulty way. The procedure—the algorithm—used to get the answer is incorrect. An incorrect procedure repeated in problem-solving is called a defective algorithm and can lead to a group of related errors called an **error pattern.**

Research into the errors students make on tests shows that error patterns show up in the work of many students. These errors can provide valuable clues to understanding why an error was made. Not only can they tell an instructor what mistaken steps the student is taking but also what mathematical concepts the student hasn't learned (or has learned incorrectly). This makes the study of error patterns a valuable remedial tool. By diagnosing a specific problem, a specific remedy can be applied.

By itself, assigning additional exercises may do no good at all because a faulty method can lead to a correct answer. If the student gets the correct answer using a faulty algorithm, it can actually reinforce the error. It may sound strange to say that a faulty algorithm can lead to a correct answer. However, take the case of a student who confuses rounding to the nearest dollar with rounding to the next higher dollar. The student's method in these examples is to change the cents digits to zeros and add one to the dollars place. This student does not understand the concept of rounding off to the nearest place and is using an incorrect method. But whenever the cents are 50 or more (50 percent of the time), the next higher dollar is the nearest dollar and the answer is right. A student who gets enough correct answers from an incorrect method is not likely to understand why the method is incorrect or to change the method used.

Let's look at a common error to show how an analysis of error patterns can help pinpoint a student's problem and suggest a remedy. The following are problems done by a student who is having difficulty adding two digit numbers.

$$\begin{array}{r}43\\+55\\\hline 98\end{array}\qquad\begin{array}{r}37\\+68\\\hline 915\end{array}\qquad\begin{array}{r}24\\+61\\\hline 85\end{array}\qquad\begin{array}{r}25\\+15\\\hline 310\end{array}$$

Since one cause of mistakes can be a lack of knowledge of basic number facts, possible errors here are checked for first. Each column of digits is added correctly. However, the student has not carried in any column. Instead, each column of digits is treated as if it were a separate problem. This leads one to think that the student does not know how to carry numbers in addition and perhaps also does not understand place value. Notice, however, that the student got half of the problems correct using a faulty algorithm and would continue to get correct answers as long as carrying was not necessary.

Basic Mathematics has tried to incorporate the basic principle of error-pattern analysis and remediation throughout the program. In the Skills Analyses, the student is called upon to solve a multiple-choice question that has its basis in various identified common error patterns. One answer is correct; the others represent different common error patterns in the basic arithmetic operations. (All of the defective algorithms in this textbook are documented in the literature or else have been personally observed by the authors while working with students.) If the student selects an incorrect answer, he or she is routed to the back of the textbook for help in understanding how to solve the problem and identifying what is going wrong. The student reviews the instruction, sees a step-by-step breakdown of how the problem is solved, and is given additional practice problems.

This material is meant to be a foundation upon which you and the student can build further insight into why he or she is having problems. Students who have formed the common error patterns covered can be given assistance with the help of the textbook in remedying these problems. Students making other errors (whether through lack of understanding, hurrying, or even carelessness) need to be helped in other ways.

Error patterns were also used in making up the tests the students take going through this program. Whenever possible, the incorrect answers to the problems in the Prechecks, Quizzes, and Goal Evaluations are the answers one would get using common defective algorithms. If you have the time, you may wish to go over the tests with students who do poorly.

To assist a student in remedying an error pattern, you may wish to try the following:

- Tell the student to look at the answer and see if there is an obvious error: a digit he or she forgot to write down, a decimal point that doesn't belong, etc. Often one can get a valuable clue by doing this.
- Ask the student to explain, in the student's own way, how the answer was derived. At this stage, just listen; don't try to correct the student's method or make any judgmental statement. Simply attempt to identify and understand the student's error in full, not correct it. If the student feels you are disparaging the method used, you are less likely to get the student to try to explain the steps used. Keep the atmosphere as relaxed as possible; anxiety can destroy the student's ability to reason about mathematics.

Using as an example the addition problems shown above, ask the student to explain the steps taken. If the student adding 37 and 68 said, "7 + 8 = 15, 3 + 6 = 9, the answer is 915," you can see that the number facts are correct. You can also see that the student is not "carrying" the 1. To check that this is really the reason why the mistakes are made, it might be a good idea to give the student several other addition problems to do which involve carrying. If there is a pattern of errors in which the student does not carry, you have good evidence that you have found where the problem is.

It is important to remember when listening to the student explain how the problem was solved that there are many different algorithms which can be used to get the same result. Often a student may have learned an unfamiliar algorithm in another school system where that is the way being taught. It can confuse a student who has learned the "right" way to solve a problem if you start to criticize that method, replacing it with another method. Before criticizing the student's method as being incorrect, take a minute to see if it is an unfamiliar, but equally valid, way of solving the problem.

When instructing the student who is using a faulty algorithm, it is important to give the conceptual basis of the algorithm first. If the student does not understand why an operation works, it is likely that the student will just learn a new series of steps by rote. In that case, the student is much more likely to get into further trouble in the future. Only after the student understands the basis of the algorithm should you give the correct algorithm.

For the instructor interested in further information on the subject of diagnostic teaching of mathematics and of error-pattern analysis, two excellent texts are: Robert B. Ashlock, *Error Patterns in Computation*, 2/e (Columbus, Ohio: Charles E. Merrill, 1976); Fredricka K. Reisman, *A Guide to the Diagnostic Teaching of Arithmetic*, 2/e (Columbus, Ohio: Charles E. Merrill, 1978).

SPEED DRILLS

The Speed Drills in this program are designed to be used as warm-up activities for the beginning of a class period. The student should be allowed about 90 seconds to do each drill. The drills can then be graded by a student monitor or by the students themselves. Students who score over 90 percent on the drills go on, while students who get lower scores repeat the drills until they do pass. The students keep an individual record of the drills passed and of their passing scores.

These Speed Drills help the students who are weak in number facts toward mastery by pinpointing the number facts they still need to learn. Each student goes along at an individual pace, depending upon the student's mastery of the facts. Students should be given as much positive reinforcement as possible as a motivation to commit these number facts to memory.

• ADDITION FACTS—SPEED DRILL 1

1 +8	1 +1	1 +5	0 +6	0 +0	3 +4
6 +5	9 +4	7 +8	6 +7	6 +8	1 +9
1 +4	5 +8	1 +0	2 +8	9 +0	3 +9
4 +0	3 +0	7 +3	2 +9	6 +6	3 +3
0 +2	9 +6	8 +9	8 +8		

• ADDITION FACTS—SPEED DRILL 2

0 +5	2 +2	1 +6	7 +1	7 +7	2 +6
8 +4	8 +3	8 +0	9 +9	7 +2	5 +7
4 +2	3 +5	0 +7	5 +5	5 +9	
4 +4	6 +4	7 +4	6 +3	1 +3	
2 +1	2 +5	3 +2	5 +4	9 +7	

• SUBTRACTION FACTS—SPEED DRILL 1

10 −9	17 −8	11 −3	1 −1	10 −4
9 −5	8 −4	16 −8	1 −0	10 −5
6 −2	14 −7	11 −5	7 −3	3 −1
9 −4	10 −2	7 −6	6 −3	4 −2
3 −3	6 −5	15 −9	14 −5	10 −1

• SUBTRACTION FACTS—SPEED DRILL 2

4 −1	13 −7	7 −2	9 −6	14 −8
16 −9	6 −1	3 −0	4 −0	12 −7
12 −4	8 −7	14 −9	8 −2	10 −3
11 −9	6 −0	15 −6	11 −6	13 −9
12 −5	11 −2	12 −9	6 −6	18 −9

• SUBTRACTION FACTS—SPEED DRILL 3

2 −0	8 −6	11 −8	16 −7	7 −5
9 −7	13 −4	10 −8	13 −5	11 −7
12 −3	15 −7	9 −0	9 −2	10 −7
5 −4	17 −9	8 −3	3 −2	9 −9
7 −4	8 −1	7 −0	9 −1	4 −4

• SUBTRACTION FACTS—SPEED DRILL 4

10 −6	5 −2	8 −0	12 −6	6 −4	8 −5	7 −7
13 −8	5 −5	7 −1	9 −3	15 −8	4 −3	
8 −8	5 −1	2 −2	5 −3	5 −0	11 −4	
13 −6	2 −1	9 −8	14 −6	12 −8	0 −0	

• MULTIPLICATION FACTS—SPEED DRILL 1

7 ×4	7 ×3	9 ×6	7 ×2	5 ×9	9 ×9
1 ×3	1 ×6	7 ×8	5 ×7	8 ×0	3 ×3
6 ×7	2 ×6	2 ×8	1 ×9	6 ×6	2 ×2
1 ×1	1 ×0	4 ×4	5 ×8	0 ×0	2 ×1
0 ×6	4 ×0	2 ×5	8 ×8		

• MULTIPLICATION FACTS—SPEED DRILL 2

0 ×5	1 ×8	5 ×4	8 ×9	6 ×3	7 ×1
9 ×0	9 ×7	0 ×2	0 ×7	6 ×5	3 ×9
3 ×4	6 ×8	9 ×4	8 ×3	1 ×5	
7 ×7	3 ×0	3 ×2	6 ×4	5 ×5	
2 ×9	4 ×2	3 ×5	8 ×4	1 ×4	

• DIVISION FACTS—SPEED DRILL 1

$8\overline{)32}$	$2\overline{)10}$	$7\overline{)63}$	$7\overline{)56}$	$9\overline{)0}$
$4\overline{)28}$	$9\overline{)18}$	$6\overline{)36}$	$1\overline{)4}$	$7\overline{)35}$
$4\overline{)24}$	$3\overline{)0}$	$5\overline{)30}$	$8\overline{)72}$	$3\overline{)12}$
$4\overline{)16}$	$4\overline{)0}$	$6\overline{)54}$	$6\overline{)30}$	$3\overline{)24}$
$5\overline{)10}$	$4\overline{)32}$			

• DIVISION FACTS—SPEED DRILL 2

$1\overline{)9}$	$5\overline{)40}$	$8\overline{)0}$	$2\overline{)18}$	$9\overline{)81}$
$6\overline{)0}$	$4\overline{)12}$	$2\overline{)14}$	$6\overline{)6}$	$8\overline{)56}$
$4\overline{)4}$	$8\overline{)16}$	$8\overline{)40}$	$5\overline{)0}$	$7\overline{)21}$
$5\overline{)5}$	$8\overline{)48}$	$2\overline{)0}$	$5\overline{)35}$	$5\overline{)15}$
$1\overline{)5}$	$9\overline{)9}$			

• DIVISION FACTS—SPEED DRILL 3

$1\overline{)8}$	$9\overline{)72}$	$8\overline{)8}$	$3\overline{)15}$	$9\overline{)27}$
$1\overline{)0}$	$5\overline{)20}$	$6\overline{)24}$	$7\overline{)0}$	$7\overline{)49}$
$7\overline{)42}$	$9\overline{)54}$	$4\overline{)36}$	$9\overline{)45}$	$3\overline{)18}$
$6\overline{)42}$	$2\overline{)2}$	$2\overline{)16}$	$3\overline{)27}$	$1\overline{)2}$
$1\overline{)1}$	$6\overline{)12}$			

• DIVISION FACTS—SPEED DRILL 4

$5\overline{)45}$	$8\overline{)64}$	$7\overline{)28}$	$3\overline{)3}$	$6\overline{)18}$
$3\overline{)6}$	$1\overline{)3}$	$6\overline{)48}$	$4\overline{)20}$	$8\overline{)24}$
$3\overline{)9}$	$1\overline{)6}$	$3\overline{)21}$	$2\overline{)6}$	$9\overline{)63}$
$4\overline{)8}$	$5\overline{)25}$	$2\overline{)4}$	$7\overline{)14}$	$1\overline{)7}$
$7\overline{)7}$	$2\overline{)12}$	$9\overline{)36}$	$2\overline{)8}$	

• **FRACTIONS-TO-DECIMALS—SPEED DRILL**

$\frac{1}{2}$ =	$\frac{9}{10}$ =	$\frac{5}{8}$ =	$\frac{2}{3}$ =
$\frac{3}{4}$ =	$\frac{1}{3}$ =	$\frac{7}{8}$ =	$\frac{1}{4}$ =
$\frac{1}{5}$ =	$\frac{1}{10}$ =	$\frac{4}{5}$ =	$\frac{1}{8}$ =
0.7 =	0.4 =	.6 =	.375 =
.125 =	.3 =	.625 =	.25 =
.875 =	.9 =	.75 =	.5 =
.1 =			

• **FRACTIONS-TO-PERCENTS—SPEED DRILL**

$\frac{1}{5}$ =	$\frac{5}{8}$ =	$\frac{4}{5}$ =	$\frac{2}{3}$ =
$\frac{1}{10}$ =	$\frac{1}{2}$ =	$\frac{7}{8}$ =	$\frac{9}{10}$ =
$\frac{1}{8}$ =	$\frac{1}{4}$ =	$\frac{3}{4}$ =	$\frac{1}{3}$ =
7% =	$33\frac{1}{3}$% =	40% =	$62\frac{1}{2}$% =
3% =	25% =	60% =	$87\frac{1}{2}$% =
2% =	$66\frac{2}{3}$% =	1% =	80% =
$12\frac{1}{2}$% =			

ANSWERS TO SPEED DRILLS

• ADDITION FACTS—SPEED DRILL 1

1	1	1	0	0	3
+8	+1	+5	+6	+0	+4
9	**2**	**6**	**6**	**0**	**7**
6	9	7	6	6	1
+5	+4	+8	+7	+8	+9
11	**13**	**15**	**13**	**14**	**10**
1	5	1	2	9	3
+4	+8	+0	+8	+0	+9
5	**13**	**1**	**10**	**9**	**12**
4	3	7	2	6	3
+0	+0	+3	+9	+6	+3
4	**3**	**10**	**11**	**12**	**6**
0	9	8	8		
+2	+6	+9	+8		
2	**15**	**17**	**16**		

• ADDITION FACTS—SPEED DRILL 2

0	2	1	7	7	2
+5	+2	+6	+1	+7	+6
5	**4**	**7**	**8**	**14**	**8**
8	8	8	9	7	5
+4	+3	+0	+9	+2	+7
12	**11**	**8**	**18**	**9**	**12**
4	3	0	5	5	
+2	+5	+7	+5	+9	
6	**8**	**7**	**10**	**14**	
4	6	7	6	1	
+4	+4	+4	+3	+3	
8	**10**	**11**	**9**	**4**	
2	2	3	5	9	
+1	+5	+2	+4	+7	
3	**7**	**5**	**9**	**16**	

• SUBTRACTION FACTS—SPEED DRILL 1

10 − 9 = **1**	17 − 8 = **9**	11 − 3 = **8**	1 − 1 = **0**	10 − 4 = **6**
9 − 5 = **4**	8 − 4 = **4**	16 − 8 = **8**	1 − 0 = **1**	10 − 5 = **5**
6 − 2 = **4**	14 − 7 = **7**	11 − 5 = **6**	7 − 3 = **4**	3 − 1 = **2**
9 − 4 = **5**	10 − 2 = **8**	7 − 6 = **1**	6 − 3 = **3**	4 − 2 = **2**
3 − 3 = **0**	6 − 5 = **1**	15 − 9 = **6**	14 − 5 = **9**	10 − 1 = **9**

• SUBTRACTION FACTS—SPEED DRILL 2

4 − 1 = **3**	13 − 7 = **6**	7 − 2 = **5**	9 − 6 = **3**	14 − 8 = **6**
16 − 9 = **7**	6 − 1 = **5**	3 − 0 = **4**	4 − 0 = **4**	12 − 7 = **5**
12 − 4 = **8**	8 − 7 = **1**	14 − 9 = **5**	8 − 2 = **6**	10 − 3 = **7**
11 − 9 = **2**	6 − 0 = **6**	15 − 6 = **9**	11 − 6 = **5**	13 − 9 = **4**
12 − 5 = **7**	11 − 2 = **9**	12 − 9 = **3**	6 − 6 = **0**	18 − 9 = **9**

• SUBTRACTION FACTS—SPEED DRILL 3

2 − 0 = **2**	8 − 6 = **2**	11 − 8 = **3**	16 − 7 = **9**	7 − 5 = **2**
9 − 7 = **2**	13 − 4 = **9**	10 − 8 = **2**	13 − 5 = **8**	11 − 7 = **4**
12 − 3 = **9**	15 − 7 = **8**	9 − 0 = **9**	9 − 2 = **7**	10 − 7 = **3**
5 − 4 = **1**	17 − 9 = **8**	8 − 3 = **5**	3 − 2 = **1**	9 − 9 = **0**
7 − 4 = **3**	8 − 1 = **7**	7 − 0 = **7**	9 − 1 = **8**	4 − 4 = **0**

• SUBTRACTION FACTS—SPEED DRILL 4

$10 - 6 = \mathbf{4}$	$5 - 2 = \mathbf{3}$	$8 - 0 = \mathbf{8}$	$12 - 6 = \mathbf{6}$	$6 - 4 = \mathbf{2}$	$8 - 5 = \mathbf{3}$	$7 - 7 = \mathbf{0}$
$13 - 8 = \mathbf{5}$	$5 - 5 = \mathbf{0}$	$7 - 1 = \mathbf{6}$	$9 - 3 = \mathbf{6}$	$15 - 8 = \mathbf{7}$	$4 - 3 = \mathbf{1}$	
$8 - 8 = \mathbf{0}$	$5 - 1 = \mathbf{4}$	$2 - 2 = \mathbf{0}$	$5 - 3 = \mathbf{2}$	$5 - 0 = \mathbf{5}$	$11 - 4 = \mathbf{7}$	
$13 - 6 = \mathbf{7}$	$2 - 1 = \mathbf{1}$	$9 - 8 = \mathbf{1}$	$14 - 6 = \mathbf{8}$	$12 - 8 = \mathbf{4}$	$0 - 0 = \mathbf{0}$	

• MULTIPLICATION FACTS—SPEED DRILL 1

$7 \times 4 = \mathbf{28}$	$7 \times 3 = \mathbf{21}$	$9 \times 6 = \mathbf{54}$	$7 \times 2 = \mathbf{14}$	$5 \times 9 = \mathbf{45}$	$9 \times 9 = \mathbf{81}$
$1 \times 3 = \mathbf{3}$	$1 \times 6 = \mathbf{6}$	$7 \times 8 = \mathbf{56}$	$5 \times 7 = \mathbf{35}$	$8 \times 0 = \mathbf{0}$	$3 \times 3 = \mathbf{9}$
$6 \times 7 = \mathbf{42}$	$2 \times 6 = \mathbf{12}$	$2 \times 8 = \mathbf{16}$	$1 \times 9 = \mathbf{9}$	$6 \times 6 = \mathbf{36}$	$2 \times 2 = \mathbf{4}$
$1 \times 1 = \mathbf{1}$	$1 \times 0 = \mathbf{0}$	$4 \times 4 = \mathbf{16}$	$5 \times 8 = \mathbf{40}$	$0 \times 0 = \mathbf{9}$	$2 \times 1 = \mathbf{2}$
$0 \times 6 = \mathbf{0}$	$4 \times 0 = \mathbf{0}$	$2 \times 5 = \mathbf{10}$	$8 \times 8 = \mathbf{64}$		

• MULTIPLICATION FACTS—SPEED DRILL 2

$\begin{array}{r}0\\ \times 5\\ \hline \mathbf{0}\end{array}$	$\begin{array}{r}1\\ \times 8\\ \hline \mathbf{8}\end{array}$	$\begin{array}{r}5\\ \times 4\\ \hline \mathbf{20}\end{array}$	$\begin{array}{r}8\\ \times 9\\ \hline \mathbf{72}\end{array}$	$\begin{array}{r}6\\ \times 3\\ \hline \mathbf{18}\end{array}$	$\begin{array}{r}7\\ \times 1\\ \hline \mathbf{7}\end{array}$
$\begin{array}{r}9\\ \times 0\\ \hline \mathbf{0}\end{array}$	$\begin{array}{r}9\\ \times 7\\ \hline \mathbf{63}\end{array}$	$\begin{array}{r}0\\ \times 2\\ \hline \mathbf{0}\end{array}$	$\begin{array}{r}0\\ \times 7\\ \hline \mathbf{0}\end{array}$	$\begin{array}{r}6\\ \times 5\\ \hline \mathbf{30}\end{array}$	$\begin{array}{r}3\\ \times 9\\ \hline \mathbf{27}\end{array}$
$\begin{array}{r}3\\ \times 4\\ \hline \mathbf{12}\end{array}$	$\begin{array}{r}6\\ \times 8\\ \hline \mathbf{48}\end{array}$	$\begin{array}{r}9\\ \times 4\\ \hline \mathbf{36}\end{array}$	$\begin{array}{r}8\\ \times 3\\ \hline \mathbf{24}\end{array}$	$\begin{array}{r}1\\ \times 5\\ \hline \mathbf{5}\end{array}$	
$\begin{array}{r}7\\ \times 7\\ \hline \mathbf{49}\end{array}$	$\begin{array}{r}3\\ \times 0\\ \hline \mathbf{0}\end{array}$	$\begin{array}{r}3\\ \times 2\\ \hline \mathbf{6}\end{array}$	$\begin{array}{r}6\\ \times 4\\ \hline \mathbf{24}\end{array}$	$\begin{array}{r}5\\ \times 5\\ \hline \mathbf{25}\end{array}$	
$\begin{array}{r}2\\ \times 9\\ \hline \mathbf{18}\end{array}$	$\begin{array}{r}4\\ \times 2\\ \hline \mathbf{8}\end{array}$	$\begin{array}{r}3\\ \times 5\\ \hline \mathbf{15}\end{array}$	$\begin{array}{r}8\\ \times 4\\ \hline \mathbf{32}\end{array}$	$\begin{array}{r}1\\ \times 4\\ \hline \mathbf{4}\end{array}$	

• DIVISION FACTS—SPEED DRILL 1

$\begin{array}{r}\mathbf{4}\\ 8\overline{)32}\end{array}$	$\begin{array}{r}\mathbf{5}\\ 2\overline{)10}\end{array}$	$\begin{array}{r}\mathbf{9}\\ 7\overline{)63}\end{array}$	$\begin{array}{r}\mathbf{8}\\ 7\overline{)56}\end{array}$	$\begin{array}{r}\mathbf{0}\\ 9\overline{)0}\end{array}$
$\begin{array}{r}\mathbf{7}\\ 4\overline{)28}\end{array}$	$\begin{array}{r}\mathbf{2}\\ 9\overline{)18}\end{array}$	$\begin{array}{r}\mathbf{6}\\ 6\overline{)36}\end{array}$	$\begin{array}{r}\mathbf{4}\\ 1\overline{)4}\end{array}$	$\begin{array}{r}\mathbf{5}\\ 7\overline{)35}\end{array}$
$\begin{array}{r}\mathbf{6}\\ 4\overline{)24}\end{array}$	$\begin{array}{r}\mathbf{0}\\ 3\overline{)0}\end{array}$	$\begin{array}{r}\mathbf{6}\\ 5\overline{)30}\end{array}$	$\begin{array}{r}\mathbf{9}\\ 8\overline{)72}\end{array}$	$\begin{array}{r}\mathbf{4}\\ 3\overline{)12}\end{array}$
$\begin{array}{r}\mathbf{4}\\ 4\overline{)16}\end{array}$	$\begin{array}{r}\mathbf{0}\\ 4\overline{)0}\end{array}$	$\begin{array}{r}\mathbf{9}\\ 6\overline{)54}\end{array}$	$\begin{array}{r}\mathbf{5}\\ 6\overline{)30}\end{array}$	$\begin{array}{r}\mathbf{8}\\ 3\overline{)24}\end{array}$
$\begin{array}{r}\mathbf{2}\\ 5\overline{)10}\end{array}$	$\begin{array}{r}\mathbf{8}\\ 4\overline{)32}\end{array}$			

• DIVISION FACTS—SPEED DRILL 2

$\begin{array}{r}9\\1\overline{)9}\end{array}$	$\begin{array}{r}8\\5\overline{)40}\end{array}$	$\begin{array}{r}0\\8\overline{)0}\end{array}$	$\begin{array}{r}9\\2\overline{)18}\end{array}$	$\begin{array}{r}9\\9\overline{)81}\end{array}$
$\begin{array}{r}0\\6\overline{)0}\end{array}$	$\begin{array}{r}3\\4\overline{)12}\end{array}$	$\begin{array}{r}7\\2\overline{)14}\end{array}$	$\begin{array}{r}1\\6\overline{)6}\end{array}$	$\begin{array}{r}7\\8\overline{)56}\end{array}$
$\begin{array}{r}1\\4\overline{)4}\end{array}$	$\begin{array}{r}2\\8\overline{)16}\end{array}$	$\begin{array}{r}5\\8\overline{)40}\end{array}$	$\begin{array}{r}0\\5\overline{)0}\end{array}$	$\begin{array}{r}3\\7\overline{)21}\end{array}$
$\begin{array}{r}1\\5\overline{)5}\end{array}$	$\begin{array}{r}6\\8\overline{)48}\end{array}$	$\begin{array}{r}0\\2\overline{)0}\end{array}$	$\begin{array}{r}7\\5\overline{)35}\end{array}$	$\begin{array}{r}3\\5\overline{)15}\end{array}$
$\begin{array}{r}5\\1\overline{)5}\end{array}$	$\begin{array}{r}1\\9\overline{)9}\end{array}$			

• DIVISION FACTS—SPEED DRILL 3

$\begin{array}{r}8\\1\overline{)8}\end{array}$	$\begin{array}{r}8\\9\overline{)72}\end{array}$	$\begin{array}{r}1\\8\overline{)8}\end{array}$	$\begin{array}{r}5\\3\overline{)15}\end{array}$	$\begin{array}{r}3\\9\overline{)27}\end{array}$
$\begin{array}{r}0\\1\overline{)0}\end{array}$	$\begin{array}{r}4\\5\overline{)20}\end{array}$	$\begin{array}{r}4\\6\overline{)24}\end{array}$	$\begin{array}{r}0\\7\overline{)0}\end{array}$	$\begin{array}{r}7\\7\overline{)49}\end{array}$
$\begin{array}{r}6\\7\overline{)42}\end{array}$	$\begin{array}{r}6\\9\overline{)54}\end{array}$	$\begin{array}{r}9\\4\overline{)36}\end{array}$	$\begin{array}{r}5\\9\overline{)45}\end{array}$	$\begin{array}{r}6\\3\overline{)18}\end{array}$
$\begin{array}{r}7\\6\overline{)42}\end{array}$	$\begin{array}{r}1\\2\overline{)2}\end{array}$	$\begin{array}{r}8\\2\overline{)16}\end{array}$	$\begin{array}{r}9\\3\overline{)27}\end{array}$	$\begin{array}{r}2\\1\overline{)2}\end{array}$
$\begin{array}{r}1\\1\overline{)1}\end{array}$	$\begin{array}{r}2\\6\overline{)12}\end{array}$			

• DIVISION FACTS—SPEED DRILL 4

$\begin{array}{r}9\\5\overline{)45}\end{array}$	$\begin{array}{r}8\\8\overline{)64}\end{array}$	$\begin{array}{r}4\\7\overline{)28}\end{array}$	$\begin{array}{r}1\\3\overline{)3}\end{array}$	$\begin{array}{r}3\\6\overline{)18}\end{array}$
$\begin{array}{r}2\\3\overline{)6}\end{array}$	$\begin{array}{r}3\\1\overline{)3}\end{array}$	$\begin{array}{r}8\\6\overline{)48}\end{array}$	$\begin{array}{r}5\\4\overline{)20}\end{array}$	$\begin{array}{r}3\\8\overline{)24}\end{array}$
$\begin{array}{r}3\\3\overline{)9}\end{array}$	$\begin{array}{r}6\\1\overline{)6}\end{array}$	$\begin{array}{r}7\\3\overline{)21}\end{array}$	$\begin{array}{r}3\\2\overline{)6}\end{array}$	$\begin{array}{r}7\\9\overline{)63}\end{array}$
$\begin{array}{r}2\\4\overline{)8}\end{array}$	$\begin{array}{r}5\\5\overline{)25}\end{array}$	$\begin{array}{r}2\\2\overline{)4}\end{array}$	$\begin{array}{r}2\\7\overline{)14}\end{array}$	$\begin{array}{r}7\\1\overline{)7}\end{array}$
$\begin{array}{r}1\\7\overline{)7}\end{array}$	$\begin{array}{r}6\\2\overline{)12}\end{array}$	$\begin{array}{r}4\\9\overline{)36}\end{array}$	$\begin{array}{r}4\\2\overline{)8}\end{array}$	

• FRACTIONS-TO-DECIMALS—SPEED DRILL

$\frac{1}{2} = 0.5$	$\frac{9}{10} = 0.9$	$\frac{5}{8} = 0.625$	$\frac{2}{3} = 0.667$
$\frac{3}{4} = .75$	$\frac{1}{3} = 0.333$	$\frac{7}{8} = 0.875$	$\frac{1}{4} = 0.25$
$\frac{1}{5} = 0.2$	$\frac{1}{10} = 0.1$	$\frac{4}{5} = 0.8$	$\frac{1}{8} = 0.125$
$0.7 = \frac{7}{10}$	$0.4 = \frac{2}{5}$	$.6 = \frac{3}{5}$	$.375 = \frac{3}{8}$
$.125 = \frac{1}{8}$	$.3 = \frac{3}{10}$	$.625 = \frac{5}{8}$	$.25 = \frac{1}{4}$
$.875 = \frac{7}{8}$	$.9 = \frac{9}{10}$	$.75 = \frac{3}{4}$	$.5 = \frac{1}{2}$
$.1 = \frac{1}{10}$			

• FRACTIONS-TO-PERCENTS—SPEED DRILL

$\frac{1}{5} = 20\%$	$\frac{5}{8} = 62\frac{1}{2}\%$	$\frac{4}{5} = 80\%$	$\frac{2}{3} = 66\frac{2}{3}\%$
$\frac{1}{10} = 10\%$	$\frac{1}{2} = 50\%$	$\frac{7}{8} = 87\frac{1}{2}\%$	$\frac{9}{10} = 90\%$
$\frac{1}{8} = 12\frac{1}{2}\%$	$\frac{1}{4} = 25\%$	$\frac{3}{4} = 75\%$	$\frac{1}{3} - 33\frac{1}{3}\%$
$7\% = \frac{7}{100}$	$33\frac{1}{3}\% = \frac{3,333}{10,000}$ or $\frac{1}{3}$	$40\% = \frac{2}{5}$	$62\frac{1}{2}\% = \frac{5}{8}$
$3\% = \frac{3}{100}$	$25\% = \frac{1}{4}$	$60\% = \frac{3}{5}$	$87\frac{1}{2}\% = \frac{7}{8}$
$2\% - \frac{1}{50}$	$66\frac{2}{3}\% = \frac{6,667}{10,000}$ or $\frac{2}{3}$	$1\% = \frac{1}{100}$	$80\% = \frac{4}{5}$
$12\frac{1}{2}\% = \frac{1}{8}$			

TEXTBOOK ANSWER KEY

• GOAL 1 • READING AND WRITING NUMBERS

Goal Practice (page 4)

1. a. millions
   b. hundred millions
   c. ones or units
   d. hundreds
2. a. 3
   b. 2
   c. 6
   d. 1
3. a. 68
   333
   841
   7,911
   8,093
   76,431
   b. 70
   814
   881
   983
   1,071
   3,641
   9,732
4. a. nine hundreds or 900
   b. nine tens or 90
   c. nine thousands or 9,000
5. a. $25,099
   b. $3,337,482
   c. $11,500

Goal Review (page 5)

1. a. ten millions
   b. ten thousands
   c. hundred thousands
   d. ones or units
2. a. 7
   b. 8
   c. 8
   d. 1
3. a. 811
   624
   314
   262
   180
   155
   99
   90
   b. 9,091
   1,601
   682
   451
   97
   33
   32
4. a. five tens or 50
   b. five ten thousands or 50,000
   c. five millions or 5,000,000
5. a. $8,754,000
   b. 2,654,000
   c. 5,489

Goal Practice (page 8)

1. six million, four hundred seven thousand, six hundred fourteen
2. nine hundred ninety-four thousand, nine hundred twenty-nine
3. forty-three million, eleven thousand, six hundred eighty-five
4. eleven thousand, five hundred
5. one hundred forty-five million, one hundred eighty-seven thousand
6. one thousand, nine hundred thirty-three

Goal Review (page 8)

1. eighty thousand, six hundred forty-three
2. nine hundred three million, forty-four
3. six million, eight hundred forty thousand, fifty-seven
4. eleven million, two hundred thousand
5. one thousand, four hundred fifty-four
6. eight million, five hundred two thousand
7. twelve million, six hundred sixty-four thousand, two hundred sixty-three
8. two hundred nine thousand, five hundred

Goal Practice (page 9)

1. 43,607
2. 5,000,403
3. 63,467,200
4. 16,040,200
5. 550,000
6. 645
7. 4,500
8. 1,700,000

Goal Review (page 10)

1. 9,041,707
2. 300,076,049
3. 7,050,040
4. 59,037
5. 100,000,000
6. 25,270,000
7. 300,000
8. 150,000
9. 695,000,000
10. 6,540

Goal Practice (page 12)

1. a. seven tenths
   b. seven thousandths
   c. seven thousandths
   d. seven hundredths
2. a. two tenths
   b. two thousandths
   c. two ten thousandths
   d. two hundredths
3. a. 2.13
   1.62
   .9
   .76349
   .007
   b. 8.3
   6.49
   4.83
   .83
   .33333

4. a. nine thousandths of a percent
   b. nine tenths of a million dollars
   c. nine billion dollars
   d. nine tenths of a percent
5. a. one ten thousandth of a percent
   b. one tenth of a million dollars
   c. one hundredth of a percent
   d. one thousandth of a percent

Goal Review (page 13)

1. a. five tenths
   b. five hundredths
   c. five thousandths
   d. five ten thousandths
2. a. six tenths
   b. six hundredths
   c. six thousandths
   d. six ten thousandths
3. a. .08
   .103
   .721
   .8
   6.007
   b. .4332
   1.28
   1.298
   4.332
   64.19
4. a. three hundredths
   b. three hundredths
   c. three ones or units
   d. three thousandths
5. a. seven tenths of a percent
   b. seven thousandths
   c. seven thousandths
   d. seven hundredths
6. a. five hundredths of a percent
   b. five thousandths of a percent

Goal Practice (page 15)

1. a. seventy and two tenths
   b. ten and three hundredths
   c. seven and five hundred forty-one thousandths
   d. three hundred ninety-three and sixty-two hundredths
2. a. point six eight nine
   b. eight point zero seven
   c. point one four four seven
   d. point nine nine nine
3. a. ninety-four and five tenths
   b. one hundred sixteen and three tenths
   c. one hundred sixty-one and two tenths
   d. one hundred eighty-one and five tenths

Goal Review (page 16)

1. a. one tenth
   b. four hundred fifty-three and eight thousandths
   c. one thousandth
   d. eight and four tenths
2. a. twenty-four point zero zero nine
   b. six hundred thirty-seven point one zero eight two
   c. point zero five
   d. seven point one
3. a. fifty-nine and eight tenths degrees
   b. seventy-six and two tenths degrees
   c. eighty-three and one tenth degrees
   d. sixty and eight tenths degrees
4. a. one and fourteen hundredths inches
   b. one and thirty-four hundredths inches
   c. thirty-three hundredths of an inch
   d. fifty-five hundredths of an inch

Goal Practice (page 17)

1. .03 or 0.03
2. .047 or 0.047
3. 6.803
4. .0022 or 0.0022
5. .06 or 0.06
6. .23 or 0.23
7. .03 or 0.03
8. .6 or 0.6
9. .43 or 0.43
10. 7.16

Goal Review (page 17)

1. .7 or 0.7
2. 3.14
3. .950 or 0.950
4. 101.1
5. .51 or 0.51
6. .17 or 0.17
7. .7% or 0.7%
8. .8% or 0.8%
9. .344 or 0.344
10. .178 or 0.178

Goal Practice (page 19)

1. a. seven eighths
   b. sixteen and one third
   c. three fourths
2. a. sixty-six and two thirds
   b. eleven and two fifths
   c. one half
3. six and three tenths of a percent
4. thirteen and nine tenths

Goal Review (page 19)

1. a. one fifth
   b. sixteen and one third
   c. three hundred and four twelfths
2. a. twenty-eight and one fourth
   b. fourteen and one half
   c. three and five sevenths
3. four fifths
4. one third

Auditing Corner (page 20)

1. a. $121,352
   b. okay
   c. okay
   d. $273,586.43
   e. okay
2. 9,683
   9,334
   8,967
   8,631
   8,421
   8,412
   8,341
   8,315
   8,309
   8,261

Business Corner (page 21)

1.

	AMOUNT	
a.	49	99
b.	300	37
c.	6	34
d.	45	79

2. decimal point

3.

a.	75	34
b.		88
c.	782	55
d.	80	00
e.	44409	00
f.	3000	33

4. separate whole number digits

Consumer Corner (page 23)

1. a. $\frac{21}{100}$
   b. Only twenty-one cents
2. a. $8\frac{23}{100}$
   b. Eight and $\frac{23}{100}$
3. a. $5\frac{00}{100}$
   b. Five and $\frac{00}{100}$
4. a. $495\frac{7}{100}$
   b. Four hundred ninety-five and $\frac{7}{100}$
5. a. $\frac{99}{100}$
   b. Only ninety-nine cents
6. a. $680\frac{13}{100}$
   b. Six hundred eighty and $\frac{13}{100}$
7. a. $3,482\frac{10}{100}$
   b. Three thousand four hundred eighty-two and $\frac{10}{100}$
8. a. $7\frac{71}{100}$
   b. Seven and $\frac{71}{100}$
9. a. $800\frac{00}{100}$
   b. Eight hundred and $\frac{00}{100}$

• GOAL 2 • ROUNDING OFF NUMBERS AND MONEY AMOUNTS

Goal Practice (page 29)

1. a. $450
   b. $100
   c. $1,170
2. a. $0
   b. $100,000
   c. $10,000
3. a. $100,000
   b. $100,000
   c. $1,600,000
4. 49,000
5. 1,500
6. $60,000,000

Goal Review (page 29)

1. a. 500
   b. 100
   c. 0
2. a. 88,000
   b. 1,000
   c. 2,000
3. a. 420,000
   b. 10,000
   c. 10,000
4. 32,000,000
5. 3,000
6. 37

Goal Practice (page 31)

1. a. $3.84
   b. $9.05
   c. $11.30
2. a. .11
   b. 3.35
   c. .11
3. $4.593
4. $45
5. $.0996
6. 12.8%

Goal Review (page 31)

1. a. $4.333
   b. $.512
   c. $23.546
2. a. 44.369
   b. .222
   c. 99.003
3. a. 6.735
   b. 6.978
   c. .201
4. $.20
5. $.12
6. 3.083%

Goal Practice (page 32)

1. a. $.34
   b. $.25
   c. $3.99
   d. $.76
2. a. $456
   b. $19,785
   c. $11,450
   d. $687
3. a. $45.78
   b. $68
   c. $7.23
   d. $200.58

Goal Review (page 33)

1. a. $1.05
   b. $2.21
   c. $8.04
   d. $.95
2. a. $15.39
   b. $8.04
   c. $804
   d. $21.35
3. a. $1.30
   b. $5
   c. $2.74
   d. $7.62

Auditing Corner (page 33)

1. a. okay
   b. okay
   c. okay
   d. okay
   e. okay
   f. okay
   g. okay
   h. okay
   i. okay
   j. okay
2. a. $8,570,000
   b. $3,990,000
   c. okay
   d. $2,380,000
   e. okay

Business Corner (page 35)

1. a. $1,038.11
   b. $246.83
   c. $740.50
   d. $649.13
   e. $1,947.40
2. a. $47.50
   b. $12.00
   c. $7.50
   d. $9.10
   e. $65.79

Consumer Corner (page 37)

1. a. $70
   b. $73
   c. $56
   d. $48
   e. $40
   f. $36

• GOAL 3 • ADDING WHOLE NUMBERS AND MONEY AMOUNTS

Goal Practice (page 42)

1. 26
2. 25
3. 26
4. 29
5. 22
6. a. 18
   b. 19
7. 39

Goal Review (page 43)

1. 23
2. 32
3. 30
4. 26
5. 23

6. 34
7. 24
8. 22

Goal Practice (page 44)

1. 267
2. 317
3. 343
4. 128
5. 52
6. 195

Goal Review (page 45)

1. 273
2. 329
3. 225
4. 278
5. 332
6. a. 78
   b. 154
   c. 56
   d. 60
   e. 120

Auditing Corner (page 46)

1. a. 230
   b. okay
   c. okay
   d. 14
   e. okay

Business Corner

Sales Reports (page 47)

1. a. 99
   b. 417
   c. 514
   d. 1,030
2. a. 230
   b. 218
   c. 136
   d. 584
3. a. 1,449
   b. 2,096
   c. 281
   d. 3,826

Inventory Reports (page 49)

1. a. 87
   b. $1,744,832
2. a. 815
   b. $288,255
3. a. 70
   b. $649,750
4. a. 1,249
   b. $314,592

Consumer Corner (page 51)

1. $1,230
2. $680
3. $737
4. a. $638
   b. $763
5. a. $7,225
   b. $7,483

Goal Practice (page 54)

1. $18.08
2. $5.97
3. $12.73
4. $8.82
5. $23.98
6. $22.54

Goal Review (page 54)

1. $12.26
2. $14.66
3. $55.81
4. $8.04
5. $13.50
6. $45.36
7. $24.02
8. $26.15
9. $10.75
10. $20.25

Goal Practice (page 56)

1. $206.26
2. $269.26
3. $148.85
4. $125.55
5. a. $66.10
   b. $143.23
   c. $209.33

Goal Review (page 58)

1. $190.45
2. $222.18
3. $119.66
4. $133.48
5. $263.10
6. $148.39
7. $56.62
8. $158.55

Goal Practice (page 60)

1. $301.05
2. $303.10
3. $257.18
4. $1,891.69
5. $5,361.67

Goal Review (page 61)

1. $180.59
2. $533.26
3. $119.73
4. $270.87
5. $697.37
6. a. $3,673.98
   b. $3,806.45
   c. $3,119.91
   d. $3,822.95
   e. $2,772.75
   f. $2,514.41

Auditing Corner (page 62)

1. a. okay
   b. okay
   c. $93.64
   d. okay
   e. okay
   f. okay
   g. okay
   h. okay
   i. okay
   j. $53.72
   k. $279.28
2. Job 1, $135.63
   Job 2, $92.12
   Job 3, $1,289.70
   Job 4, $342.32

Business Corner (page 63)

1. $89,703.00
2. $11,490.56
3. $253,292.87
4. $696,564.68
5. Assets $229,351.28
   Liabilities $82,138.95
6. a. $91,835.91
   b. $110,823.63
   c. $110,608.78

Consumer Corner (page 65)

1. $990.36
2. $10,653.04
3. $382.93
4. $5,879.37
5. $7,249.26
6. $10,417.73
7. a. $8,328.74
   b. $9,685.26

• WORD PROBLEM STRATEGIES 1 (Page 67)

1. Known: 3 items; $6.95, $4.50, $1.59 cost of each item
   Unknown: Total cost

2. Known: $200 beginning balance; $54.50 ending balance
Unknown: Amount paid out
3. Known: 75 cartons; 10 boxes each carton
Unknown: Total boxes on hand
4. Known: $20 paid; $9.45 purchase
Unknown: Amount of change
5. Known: 140 copies; $.03 a page
Unknown: Total cost
6. Known: $2.50 cost; 5 pounds
Unknown: Cost of 1 pound
7. Known: 1,000 runners entered; 500 runners finished
Unknown: Percent of runners finished
8. Known: 30 checks; $.10 service charge per check
Unknown: Amount of service charge
9. Known: $1,000 monthly earnings; $750 monthly paycheck
Unknown: Amount of deductions
10. Known: 100 miles robot walked; 5 miles each recharging
Unknown: Number of times robot must be recharged

• GOAL 4 • SUBTRACTING WHOLE NUMBERS AND MONEY AMOUNTS

Goal Practice (page 71)

1. 19
2. 6
3. 8
4. 7
5. 8
6. 63
7. 77
8. 48
9. 58
10. 35

Goal Review (page 72)

1. 14
2. 51
3. 18
4. 19
5. 18
6. 49
7. 68
8. 38
9. 71
10. 72

Auditing Corner (page 72)

1. a. okay
b. okay
c. okay
d. okay
e. 85
f. 70
g. 69
h. 61
i. 57
j. 50
k. 110
l. 90

Business Corner (page 74)

1. a. −$2,000
b. +$1,395
c. −$1,100
d. +$404
e. +$2,251
f. +$19,981
g. +$1,577
h. −$520

Consumer Corner (page 75)

1. a. 5.5
b. 2.5
c. 9.7
d. 7.1
e. 6.9

Goal Practice (page 77)

1. $67.23
2. $79.75
3. $1.01
4. $33.03
5. $.82
6. $227.05
7. $475.32
8. $2.02

Goal Review (page 78)

1. $1,451.50
2. $11.84
3. $26.02
4. $88.89
5. $.41
6. $33.60
7. $19.88
8. $1,395.37
9. $114.35
10. $2.22

Goal Practice (page 78)

1. 14
2. 55
3. $.97
4. $1.77
5. $2.74
6. $5.25
7. $6.65
8. $.65
9. $7.23
10. $.48

Goal Review (page 79)

1. 11
2. 46
3. 29
4. $.36
5. $2.26
6. $.55
7. $6.90
8. $.68
9. 37
10. $125

Goal Practice (page 80)

1. $36.59
2. $167.89
3. $1,870.76
4. $330
5. $576.50
6. $524.50

Goal Review (page 81)

1. $13.58
2. $402.83
3. $38.29
4. $367.31
5. $21.79
6. $27,715
7. $39,272
8. $73,530
9. 3,676
10. 3,464

Auditing Corner (page 82)

1. $3.50
2. okay
3. $43
4. okay
5. okay
6. okay
7. okay
8. $5.50
9. okay
10. okay

Business Corner

Accounts Payable (page 83)

a. $13,801.97
b. $9,801.97

c. $17,219.32
d. $12,857.82
e. $9,467.55

Accounts Receivable (page 85)

a. $167.75
b. $251.05
c. $143.85
d. $424.45
e. $324.45
f. $224.45
g. $163.90

Consumer Corner (page 86)

1.

NO. 309 $ 12.30
DATE 3/10 19 —
TO Nelson Cleaners
FOR Dry cleaning

	DOLLARS	CENTS
BALANCE	643	70
AMT. DEPOSITED	0	
TOTAL	643	70
AMT. THIS CHECK	12	30
BALANCE	631	40

NO. 311 $ 111.50
DATE 3/13 19 —
TO Wmp Insurance
FOR Insurance

	DOLLARS	CENTS
BALANCE	422	60
AMT. DEPOSITED	415	20
TOTAL	837	80
AMT. THIS CHECK	111	50
BALANCE	726	30

NO. 313 $ 115.90
DATE 3/19 19 —
TO Luke & Sons
FOR TV repair

	DOLLARS	CENTS
BALANCE	725	11
AMT. DEPOSITED	0	
TOTAL	725	11
AMT. THIS CHECK	115	90
BALANCE	609	21

NO. 310 $ 208.80
DATE 3/11 19
TO Midland Bank
FOR Mortgage

	DOLLARS	CENTS
BALANCE	631	40
AMT. DEPOSITED	0	
TOTAL	631	40
AMT. THIS CHECK	208	80
BALANCE	422	60

NO. 312 $ 36.19
DATE 3/13 19 —
TO Dawn's Gifts
FOR Gift

	DOLLARS	CENTS
BALANCE	726	30
AMT. DEPOSITED	35	00
TOTAL	761	30
AMT. THIS CHECK	36	19
BALANCE	725	11

2.

NO. 114 $ 50.03
DATE 6/5 19 —
TO Central Bank
FOR Loan

	DOLLARS	CENTS
BALANCE	1479	13
AMT. DEPOSITED	0	
TOTAL	1479	13
AMT. THIS CHECK	50	03
BALANCE	1429	10

NO. 116 $ 9.27
DATE 6/9 19 —
TO Audio Exchange
FOR Stereo needle

	DOLLARS	CENTS
BALANCE	1847	31
AMT. DEPOSITED	0	
TOTAL	1847	31
AMT. THIS CHECK	9	27
BALANCE	1838	04

NO. 115 $ 13.29
DATE 6/7 19 —
TO Western Phone
FOR Phone bill

	DOLLARS	CENTS
BALANCE	1429	10
AMT. DEPOSITED	431	50
TOTAL	1860	60
AMT. THIS CHECK	13	29
BALANCE	1847	31

• GOAL 5 • MULTIPLYING WHOLE NUMBERS AND MONEY AMOUNTS

Goal Practice (page 91)

1. 342
2. 252
3. 399
4. 96
5. 168
6. 384
7. 64
8. 125

Goal Review (page 92)

1. 736
2. 174
3. 644
4. 585
5. 351
6. 135
7. 108
8. 210
9. $470

Goal Practice (page 93)

1. 3,170
2. 35,000
3. 1,800
4. 1,500
5. $560
6. 3,500

Goal Review (page 93)

1. 360
2. 4,200
3. 379,000
4. 1,200
5. 3,200
6. 450,000
7. 4,500
8. $1,200
9. $2,100
10. $300,000

Auditing Corner (page 93)

1. a. okay
   b. $141
   c. okay
   d. $268
   e. okay
   f. okay
2. a. okay
   b. 168
   c. okay
   d. okay
   e. 304
   f. okay
   g. okay
   h. 280
   i. okay

Business Corner (page 95)

1. a. 13
   b. 32
   c. 20
   d. 51
   e. 7

2. a. 140
   b. 120
   c. 24
   d. 750
   e. 540

Consumer Corner (page 96)

1. a. $538
   b. $485
   c. $593
2. a. $625
   b. $593
   c. $434
   d. $214
3. a. $1,550
   b. $840
   c. $880

Goal Practice (page 99)

1. $41.02
2. $44.73
3. $48.96
4. $224.94
5. $43.50
6. $38.80
7. $10.89
8. $79.65

Goal Review (page 100)

1. $24.71
2. $29.96
3. $31.55
4. $32.41
5. $53.28
6. $71.40
7. $7.56
8. $44.00
9. $4.36
10. $9.72

Goal Practice (page 101)

1. 2,752
2. 5,715
3. $209.92
4. $283.53
5. $136.08
6. 1,540
7. 648
8. $285.20

Goal Review (page 102)

1. 3,915
2. 1,564
3. $226.92
4. $337.44
5. $502.86
6. 528
7. 462
8. 14,445
9. 1,650
10. $2,280

Goal Practice (page 103)

1. 286,740
2. 983,052
3. $4,410
4. $228,904
5. $197,232
6. 75,555
7. $2,383,424
8. 4,495

Goal Review (page 104)

1. 56,682
2. 103,974
3. 869,130
4. 805,787
5. $41,405
6. $23,625
7. $21,600
8. $700,000
9. $2,695
10. $6,237

Goal Practice (page 106)

1. 250,920
2. 484,812
3. 207,672
4. 360
5. 177,633
6. 83,720
7. $146,640
8. $22,032

Goal Review (page 107)

1. 121,905
2. 306,432
3. 286,300
4. 136,080
5. 57,783
6. $10,800
7. $7,756.80
8. 1,050
9. Week: 72,000
   Year: 3,456,000
10. $182,500

Auditing Corner (page 108)

1. okay
2. okay
3. $5,647.50
4. $4,194.75
5. okay
6. okay
7. $3,871.75
8. okay
9. okay
10. $9,851.50
11. okay

Business Corner (page 109)

1. a. $22.95
   b. $10.47
   c. $14.10
   d. $21.60
   e. $100.12
2. a. $442.00
   b. $288.60
   c. $111.00
   d. $124.25
   e. $965.85
3. a. $752.50
   b. $564.00
   c. $114.60
   d. $75.60
   e. $22.50
   f. $1,529.20

Consumer Corner (page 112)

1. a. $256,696
   b. $293,246.20
   c. $209,989.60
   d. $250,735
   e. $233,036.80
2. a. $258,756.80
   b. $316,244

• GOAL 6 • DIVIDING WHOLE NUMBERS AND MONEY AMOUNTS

Goal Practice (page 116)

1. 448 ÷ 50
2. 42 ÷ 6
3. 72 ÷ 8
4. $48 ÷ 6
5. 70 ÷ 10
6. 480 ÷ 6
7. 32 ÷ 8
8. $100,000 ÷ 4

Goal Practice (page 117)

1. a. 1
   b. 0
   c. Impossible
   d. 16
2. a. 0
   b. 9
   c. Impossible
   d. 1

3. a. 0
   b. 1
   c. Impossible
   d. 7
4. a. Impossible
   b. 0
   c. Impossible
   d. 1
5. a. 0
   b. 2
   c. Impossible
   d. 1

Goal Practice (page 121)

1. 97
2. 85
3. 284
4. a. 77
   b. 88
   c. 536

Goal Review (page 121)

1. 147
2. 198
3. 631
4. 457
5. 218
6. 43
7. $45
8. 52
9. 391
10. 1,352

Goal Practice (page 123)

1. $19.54
2. $9.92
3. $16.50
4. a. $14.43
   b. $17.78
   c. $15.72

Goal Review (page 123)

1. $78.03
2. $19.31
3. $9.46
4. $4.68
5. $5.63
6. $14.14
7. $16.27
8. $.35
9. $2.34
10. $.29

Goal Practice (page 125)

1. $20.71
2. 65 R1
3. 6 R4
4. $1,042.07
5. 381 R1
6. 7 hours

Goal Review (page 125)

1. a. 27 R2
   b. $6.88
   c. 500
   d. 412
   e. 525
2. a. 6 days; 9 hr
   b. 5 days; 11 hr
   c. 4 days; 13 hr
   d. 5 days; 10.2 hr

Auditing Corner (page 127)

1. a. okay
   b. $.39
   c. okay
   d. okay
   e. okay
2. a. $35.56
   b. okay
   c. okay
   d. okay
   e. $25.00

Business Corner (page 128)

1. a. $3,300
   b. $330
   c. $290
2. a. $98,000
   b. $170,000
   c. $4,833.33
3. $39,000
4. $34,225
5. $4,536.11
6. $14,505

Consumer Corner

Unit Prices (page 130)

1. a. $5.49
   b. $1.95
   c. $4.49
   d. $2.65

Loss Leaders and Sales Items (page 131)

	Grocerama	Foodland
1. a.	$.20√	$.35
b.	$.98	$.96√
c.	$.25√	$.30

The Giant Economy Size (page 132)

1. a. $.18√
   b. $.20
   c. $.69
   d. $.52√
   e. $1.03√
   f. $1.25

Goal Practice (page 136)

1. 51 R12
2. 73 R35
3. 246 R31
4. a. 152 R12
   b. 215 R1
   c. 161 R25

Goal Review (page 136)

1. 168 R12
2. 133 R24
3. 95 R14
4. 508 R20
5. 866 R38
6. 3,909 R5
7. 3,658 R6
8. 60 R37
9. 93 R32
10. 511 R33

Goal Practice (page 138)

1. $1.63
2. $12.08
3. $6.02
4. a. $.36
   b. $.41
   c. $.38

Goal Review (page 138)

1. $1.42
2. $13.49
3. 88 R395
4. $5.32
5. $9.02
6. 126 R29 or 126.25
7. 48
8. $3.61
9. $.45
10. 98 R118 or 98.94

Goal Practice (page 139)

1. 51 R6
2. 60 R51
3. 44 R1
4. a. okay
   b. okay
   c. 15

Goal Review (page 140)

1. okay
2. okay
3. okay
4. 22 R8
5. 98 R2

Auditing Corner (page 140)

1. okay
2. okay
3. okay
4. $198.65
5. okay
6. okay
7. $230.01
8. okay
9. $750.25
10. okay

Business Corner

Unit Costs (page 142)

1. a. $.12; $180
   b. $.03; $384
   c. $.01; $660
   d. $.75; $712.50

Average Costs (page 143)

1. a. $1.77; $15.93
   b. $2.67; $34.71
   c. $17.64; $105.84
   d. $1.89; $28.35

Merchandise Turnover Rate (page 144)

1. a. 11
   b. 12
   c. 8

Consumer Corner (page 145)

1. a. 29 mpg
   b. yes
2. a. 31 mpg
   b. yes
3. a. 25 mpg
   b. no
4. a. 30 mpg
   b. yes
5. a. 46 mpg
   b. 21 mpg

• WORD PROBLEM STRATEGIES 2 (Page 147)

1. Known: $2,500 loan; 11 percent interest; 1 year
   Unknown: Amount of interest
   Words: How much is $2,500 times 11 percent times 1 year?
   Formula:

Int. = Prin. × Rate × Time
? = $2,500 × 11% × 1

2. Known: Sales for 3 weeks of $8,495, $7,954, $10,903
   Unknown: Total sales
   Words: How much is $8,495 plus $7,954 plus $10,903?
   Formula:

Tot. Sales = Wk. 1 Sales + Wk. 2 Sales + Wk. 3 Sales
? = $8,495 + $7,954 + $10,903

3. Known: $2,000 raise; $10,000 annual salary
   Unknown: Percent of increase in salary
   Words: What percent is $2,000 divided by $10,000?
   Formula:

% Incr. = Raise ÷ Salary
? = $2,000 ÷ $10,000

4. Known: Monthly order of 75 machines; 20% cut
   Unknown: How many fewer machines are ordered a month?
   Words: How much is 75 times 20%?
   Formula:

Decrease = Prev. Order × % Cut
? = 75 × 20%

5. Known: 24¢ a day; 19 days overdue
   Unknown: Amount charged
   Words: How much is $.24 times 19 days?
   Formula:

Amt. Due = Daily Charge × Days
? = $.24 × 19

6. Known: 31 hours of overtime; $8.50 overtime hourly rate
   Unknown: Overtime earnings
   Words: How much is 31 hours times $8.50?
   Formula:

Overtime = Overtime Hours × Rate
? = 31 × $8.50

7. Known: $2,100 amount borrowed; 15% interest; 1 year
   Unknown: Interest charges
   Words: How much is $2,100 times 15% times 1 year?
   Formula:

Int. = Prin. × Rate × Time
? = $2,100 × 15% × 1

8. Known: 13-minute call; $2.38 for the first 3 minutes; 90¢ each additional minute
   Unknown: Total charge
   Words: What is the charge for the additional minutes added to the charge for the first 3 minutes?
   Formula:

Overtime Charge = Overtime × Overtime Rate
? = 10 min. × $.90
Total Charge = Charge for Call + Overtime Charge
? = $2.38 + $9

9. Known: $15,000 policy; costs $11.55 per thousand
   Unknown: Amount paid in premiums
   Words: How much is $15,000 divided by 1,000 times $11.55?
   Formula:

Premium = No. of $1,000s × Cost per $1,000
? = 15 × $11.55

10. Known: $2,480.53 taxes paid; $2,768 actual taxes due
    Unknown: Taxes still owed
    Words: How much is $2,768 minus $2,480.53?
    Formula:

Taxes Owed = Taxes Due − Taxes Paid
? = $2,768 − $2,480.53

• GOAL 7 • PLACING THE DECIMAL POINT

Goal Practice (page 150)

1. thousandths
2. hundredths
3. thousandths
4. ten thousandths
5. tenths
6. hundred thousandths
7. thousandths
8. hundredths
9. tenths

Goal Review (page 151)

1. thousandths
2. tenths
3. ten thousandths
4. hundredths
5. tenths
6. hundred thousandths
7. hundredths
8. thousandths
9. tenths
10. tenths

Goal Practice (page 153)

1. 698.51
2. 61.762
3. 913.692
4. 186.1 miles
5. 9.58 inches
6. .80 gram

Goal Review (page 153)

1. 26
2. 73.12
3. 379.13
4. 88.2902
5. 13.23662
6. 10.2 metric tons
7. 171.63 kilograms
8. $8.36

Goal Practice (page 155)

1. .8
2. 47.8
3. 2.523
4. 2.5
5. 128.7 miles
6. .06 inch
7. 2.878 ounces
8. 5.133 thousand

Goal Review (page 155)

1. .922
2. 3,692.824
3. .2313
4. .78
5. 1.76
6. .574 minute
7. .179 gram
8. 3.07 million
9. 3.465 ounces
10. $279.93

Auditing Corner (page 156)

1. a. okay
   b. okay
   c. okay
   d. +7.37
2. a. −.488
   b. +12.606
   c. +.032
   d. okay
   e. okay
3. a. okay
   b. +1.098
   c. +188.685
   d. −51.802

Business Corner (page 158)

1. a. $.261
   b. $1.288
   c. $.56
   d. $.485
2. a. $.3975
   b. $.1375
   c. $1.18
3. $.059
4. $.733

Consumer Corner (page 159)

1. $206.22
2. $188.22
3. $402.58
4. $175.86
5. $287.21
6. $391.59
7. $344.30
8. $475.37
9. $254.28
10. $259.75

Goal Practice (page 163)

1. a. 2.28
   b. .0152
   c. 3,620.0
   d. 958.50
2. a. $156.34
   b. $172.06
3. $2,611.46
4. $390,750
5. $43,056.75

Goal Review (page 163)

1. .0155
2. .473
3. .78
4. .05
5. .005
6. $25.50
7. $31.88
8. $850

Goal Practice (page 165)

1. 14.44
2. 6.47
3. 3.26
4. 35.32 mpg
5. $3.85
6. $12.98

Goal Review (page 166)

1. .24
2. .7
3. 3,620
4. 7
5. 15.3
6. 2.16
7. $82
8. $130.57

Auditing Corner (page 166)

1. 15.022
2. 85.000
3. okay
4. 68.482
5. okay
6. 68.126
7. 13.149
8. okay
9. 8.961
10. okay

Business Corner

Bonds (page 167)

1. a. $91,750
   b. $236,500
   c. $241,100
   d. $56,125
   e. $54,250

Promissory Notes (page 168)

1. a. $400
   b. $900
   c. $665
   d. $437.50

Consumer Corner (page 171)

1. a. 5 hundred cubic feet
   b. $5.87
   c. $30.98
2. a. 259 kwh
   b. $32.77
   c. 7 hundred cubic feet
   d. $8.08
   e. $40.85
3. a. $1.92
   b. $.10246

• GOAL 8 • WRITING COMMON FRACTIONS, DECIMAL FRACTIONS, AND PERCENTS

Goal Practice (page 176)

1. a. $\frac{2}{4}$
   b. $\frac{1}{3}$
   c. $\frac{3}{5}$
   d. $\frac{9}{16}$
   e. $\frac{1}{8}$
   f. $\frac{2}{6}$
2. a. $\frac{7}{10}$
   b. $\frac{\$60}{\$1,320}$
   c. $\frac{25}{30}$
   d. $\frac{\$4,000}{\$18,000}$
   e. $\frac{300}{800}$

Goal Review (page 176)

1. a. $\frac{1}{4}$
   b. $\frac{2}{3}$
   c. $\frac{4}{5}$
   d. $\frac{5}{16}$
   e. $\frac{3}{8}$
   f. $\frac{3}{6}$
2. a. $\frac{7,000}{9,000}$
   b. $\frac{12}{40}$
   c. $\frac{15}{155}$
   d. $\frac{38}{265}$

Goal Practice (page 178)

1. $\frac{70}{90}$
2. $\frac{18}{162}$
3. $\frac{236}{180}$
4. $\frac{\$12,500}{\$37,000}$
5. $\frac{\$6,000,000}{\$2,000,000}$
6. $\frac{\$28,000}{\$36,000}$
7. $\frac{300}{425}$
8. $\frac{25,000}{23,000}$
9. $\frac{\$3,900,000}{\$4,000,000}$
10. $\frac{35,000}{45,000}$

Goal Review (page 179)

1. $\frac{437}{110}$
2. $\frac{3,400}{1,900}$
3. $\frac{\$79,000}{\$237,000}$
4. $\frac{933}{363}$
5. $\frac{90}{3,600}$
6. $\frac{17}{14}$
7. $\frac{\$1,800,000}{\$2,600,000}$
8. $\frac{\$8,500}{\$4,200}$
9. $\frac{130}{180}$
10. $\frac{24,000}{28,000}$

Goal Practice (page 181)

1. a. 16 out of 100
   b. 75 out of 1,000
   c. 77 out of 10,000
2. a. $\frac{3}{10}$
   b. $\frac{6}{1,000}$
   c. $\frac{7}{100}$
3. a. $\frac{\$300}{\$1,000} = .300$
   b. $\frac{\$4}{\$100} = .04$
   c. $\frac{\$45,000}{\$100,000} = .45000$

Goal Review (page 181)

1. a. 116 out of 1,000
   b. 11 out of 100
   c. 99 out of 1,000
2. a. $\frac{2}{100}$
   b. $\frac{776}{10,000}$
   c. $\frac{119}{1,000}$
3. a. $\frac{19}{100} = .19$
   b. $\frac{18}{100} = .18$
   c. $\frac{\$850}{\$1,000} = .850$
   d. $\frac{\$4}{\$1,000} = .004$

Goal Practice (page 183)

1. a. expenses
   b. revenue
   c. unclear
   d. last year's profit
2. a. 15
   b. $2\frac{1}{2}$
   c. 106
   d. 4 cents or $.04

Goal Review (page 184)

1. a. income
   b. income
   c. unclear
   d. first prize money
2. a. 14
   b. 7
   c. $12\frac{1}{2}$ cents or $\$.12\frac{1}{2}$
   d. 55 cents or $.55
3. b
4. c

Auditing Corner (page 185)

1. a. okay
   b. okay
   c. okay
   d. $\frac{5,473}{5,969}$
   e. okay
2. a. okay
   b. okay
   c. 15.7%
   d. okay
   e. 14.2%

Business Corner (page 186)

1. 67%
2. 49%
3. $\frac{5,000}{35,000}$
4. .80000
5. .4750
6. $\frac{6}{14}$
7. $\frac{35}{49}$
8. 4%
9. .30
10. .450

Consumer Corner (page 187)

1. $\frac{53}{59}$
2. 33%
3. .2
4. $\frac{312}{900}$ or $\frac{\$3.12}{\$9.00}$
5. 14%
6. $\frac{72}{79}$
7. $\frac{68}{119}$ or $\frac{\$.68}{\$1.19}$
8. .23000
9. .28000
10. 9%

• GOAL 9 • RENAMING COMMON FRACTIONS, DECIMAL FRACTIONS, AND PERCENTS

Goal Practice (page 191)

1. a. .125
   b. .75
   c. .6667
   d. .60
   e. .625
2. a. $\frac{6}{24}$ = .25
   b. $\frac{28}{140}$ = .20
   c. $\frac{16}{96}$ = .1667
   d. $\frac{16}{100}$ = .16
   e. $\frac{55}{840}$ = .0655
   f. $\frac{35}{50}$ = .70

Goal Review (page 192)

1. a. .25
   b. .50
   c. .80
   d. .875
   e. .20
2. a. $\frac{210}{540}$ = .3889
   b. $\frac{475}{1,550}$ = .3065
   c. $\frac{15}{32}$ = .4668
   d. $\frac{97}{140}$ = .6929
   e. $\frac{16}{100}$ = .16

Goal Practice (page 193)

1. a. 14%
   b. 3.5%
   c. 12.5%
   d. 137.5%
   e. 275%
2. a. 25%
   b. 28.5%
   c. 19.25%
   d. 66.67%
   e. 37.5%

Goal Review (page 194)

1. a. 11.11%
   b. 14.29%
   c. 75%
   d. 387.5%
   e. 6.25%
2. a. 14.5%
   b. 12.7%
   c. 4.37%
   d. 42.75%
   e. 7.5%

Goal Practice (page 195)

1. a. 1.08
   b. 30%
   c. .0107
   d. $.37\frac{1}{2}$ or .375
   e. 4%
2. a. .459
   b. 13.3%
   c. 92%
   d. 73.5%
   e. .75

Goal Review (page 195)

1. a. $.01\frac{1}{2}$ or .015
   b. 307%
   c. $.13\frac{3}{4}$ or .1375
   d. $37\frac{1}{2}$% or 37.5%
   e. .0456
2. a. 49.6%
   b. .25
   c. .852
   d. 35%
   e. .3135

Goal Practice (page 197)

1. a. $\frac{1}{4}$ = .25 = 25%
   b. $\frac{2}{3}$ = .6667 = 66.67%
   c. $\frac{3}{5}$ = .60 = 60%
   d. $\frac{9}{16}$ = .5625 = 56.25%
   e. $\frac{3}{8}$ = .375 = 37.5%
   f. $\frac{2}{6}$ = .3333 = 33.33%
2. a. $\frac{\$75}{\$300}$ = .25 = 25%
   b. $\frac{9}{15}$ = .60 = 60%
   c. $\frac{\$35}{\$70}$ = .50 = 50%
   d. $\frac{18}{24}$ = .75 = 75%
   e. $\frac{\$250}{\$350}$ = .7143 = 71.43%

Goal Review (page 198)

1. a. $\frac{3}{4}$ = .75 = 75%
   b. $\frac{1}{3}$ = .3333 = 33.33%
   c. $\frac{2}{5}$ = .40 = 40%
   d. $\frac{8}{16}$ = .50 = 50%
   e. $\frac{7}{8}$ = .875 = 87.5%
   f. $\frac{5}{6}$ = .8333 = 83.33%
2. a. $\frac{6}{48}$ = .125 = 12.5%
   b. $\frac{24}{48}$ = .50 = 50%
   c. $\frac{16}{40}$ = .40 = 40%
   d. $\frac{30}{75}$ = .40 = 40%

Auditing Corner (page 199)

1. a. 42.25
   b. okay
   c. okay
   d. 35.75
   e. okay
2. a. okay
   b. .07
   c. okay
   d. okay
   e. .1713

Business Corner (page 200)

1. a. $\frac{14,000}{30,000}$ = .4667 = 46.67%
   b. $\frac{6,500}{30,000}$ = .2167 = 21.67%
   c. $\frac{3,900}{30,000}$ = .13 = 13%
   d. $\frac{17,500}{30,000}$ = .5833 = 58.33%
   e. $\frac{4,900}{7,000}$ = .70 = 70%
2. a. 1.58%
   b. 10.86%
   c. 54.62%

d. 70.50%
e. 83.70%

Consumer Corner (page 202)

1. 6.5% or $6\frac{1}{2}$%
2. 1.75%
3. 12.5%
4. .75%

• WORD PROBLEM STRATEGIES 3 (Page 203)

1. Known: $\frac{2}{5}$
   Unknown: Percent
   Relationship: Renaming
2. Known: $30,000 budgeted; $25,040 spent
   Unknown: Amount still left
   Relationship: Subtracting
3. Known: $50,000 profit; 10 investors
   Unknown: Each investor's profit
   Relationship: Dividing
4. Known: 65 reams, 100 reams, 90 reams
   Unknown: Total reams to order
   Relationship: Adding
5. Known: 120 rolls; $3.50 per roll
   Unknown: Money earned
   Relationship: Multiplying
6. Known: $690 cost; 3 friends
   Unknown: How much each would contribute
   Relationship: Dividing
7. Known: 13,545 miles driven; 21 miles to the gallon
   Unknown: Amount of gas used
   Relationship: Dividing
8. Known: $6,500 value of car; 22% loss of value
   Unknown: How much the value went down
   Relationship: Multiplying
9. Known: $1,857.97 damage; first $250 paid by insured
   Unknown: Amount to be paid by insurance company
   Relationship: Subtracting
10. Known: Monthly charges of $725 rent, $75 cleaning fees, $273.80 utilities
    Unknown: Total monthly charges
    Relationship: Adding

• GOAL 10 • WORKING WITH TIME

Goal Practice (page 207)

1. a. 191 min
   b. 342 min
   c. 139 min
2. a. 6 hr 19 min
   b. 2 hr 52 min
   c. 13 hr 31 min
3. 130 min
4. 219 min
5. 84 min
6. 4 hr 15 min

Goal Review (page 207)

1. a. 451 min
   b. 243 min
   c. 529 min
2. a. 3 hr 49 min
   b. 11 hr 59 min
   c. 1 hr 51 min
3. 4 hr 6 min
4. 4 hr 49 min
5. 8 hr 2 min
6. 1 hr 29 min

Goal Practice (page 209)

1. 4 hr 35 min
2. 12 hr 46 min
3. 21 hr 41 min
4. 1 hr 38 min
5. 14 hr 40 min
6. 2 hr 55 min
7. 5 hr 25 min
8. 5 hr 15 min
9. 1 hr 15 min
10. 20 min

Goal Review (page 210)

1. 2 hr
2. 6 hr 21 min
3. 7 hr 47 min
4. 13 hr 27 min
5. 7 hr 44 min
6. 25 hr 38 min
7. 4 hr 10 min
8. 8 min
9. 50 min
10. 20 min

Goal Practice (page 213)

1. 5 hr 36 min
2. 5 hr 6 min
3. 1 hr 56 min
4. a. 1.5 min
   b. 1 min
   c. 1 hr 54 min

Goal Review (page 214)

1. 4 hr 10 min
2. 1 hr 29 min
3. 7 hr 35 min
4. $10.23
5. 18 hr 25 min
6. 2 hr 10 min
7. 1 hr 39 min
8. 4 hr 30 min
9. 2 hr 53 min
10. $31.41

Auditing Corner (page 214)

1. a. okay
   b. $31.50
   c. $131.99
   d. $109.08
   e. $15.84
   f. $302.16
2. a. okay
   b. $26.25
   c. $44.31
   d. okay
   e. okay
   f. okay
   g. $214.06

Business Corner (page 216)

1. $54.35
2. $83
3. $70.50
4. a. $85.50
   b. $56.25
   c. $75

Consumer Corner

Time Zones (page 218)

1. a. 5 hr 45 min
   b. 2 hr 27 min
   c. 1 hr 46 min
   d. 1 hr 31 min
   e. 4 hr 20 min
   f. 1 hr 56 min
   g. 4 hr 10 min

Military Time (page 219)

1. 9:30 A.M.
2. 6:30 P.M.
3. 10:00 A.M.
4. 1:30 P.M.
5. 7:00 P.M.
6. 10:00 A.M.
7. 2:00 P.M.
8. 7:30 P.M.
9. 10:30 A.M.
10. 6:00 P.M.

Goal Practice (page 223)

1. 29 days
2. 31 days
3. 30 days
4. 183 days
5. $2\frac{1}{2}$ years
6. 13 weeks
7. July 1
8. 208

Goal Review (page 223)

1. 31 days
2. 31 days
3. no
4. no
5. January, March, May, July, August, October, December
6. July 7
7. 60
8. March 7
9. 1980
10. July 15

Goal Practice (page 226)

1. a. July 17
b. July 8
c. August 6
d. August 16
e. July 31
2. 79 days
3. 1 year, 8 months
4. March 1, 1985
5. 11 months
6. July 14

Goal Review (page 227)

1. July 23
2. May 13
3. November 4
4. September 18
5. June 30
6. November 3
7. 2 years, 2 months
8. November 5, 1986

Auditing Corner (page 227)

1. a. May 1–May 7
b. May 8–May 24
c. May 25–May 31
d. June 1–June 15
e. June 16–July 3
f. July 4–July 11
g. July 12–July 21
2. a. okay
b. July 19
c. August 5
d. okay
e. okay
f. September 13

Business Corner

Loan Due Dates (page 228)

1. a. August 2
b. November 7
c. November 28
d. September 28
e. December 7

Cash Discount Periods (page 229)

1. a. a. September 2
b. August 13
b. a. December 18
b. 4% discount October 29; 3% discount November 8
c. a. November 27
b. November 17
d. a. August 3
b. 5% by June 14; 2% by July 4
e. a. April 6
b. 5% by March 17; 2% by March 27

Consumer Corner (page 231)

1. $168.06
2. $92.13
3. $361.81

• GOAL 11 • IDENTIFYING U.S. CUSTOMARY UNITS

Goal Practice (page 234)

1. a. pound
b. Fahrenheit
c. square yard
d. quart
e. ton
f. square foot
g. mile
h. yard
i. gallon
j. pint
k. foot
l. ounce
2. a. weight
b. temperature
c. area
d. capacity
e. weight
f. area
g. length
h. length
i. capacity
j. capacity
k. length
l. weight
3. a. square yard
b. ounce
c. degree Fahrenheit
d. mile

Goal Practice (page 236)

1. a. oz
b. sq ft
c. fl oz
d. T
e. °F
2. a. tons
b. pint
c. square yards
d. pounds
e. degrees Fahrenheit

Goal Review (page 236)

1. a. pt
b. sq ft
2. a. gal
b. sq yd
3. a. fl oz
b. mi
4. a. T
b. qt
5. a. oz
b. lb
6. yd
7. in.
8. lb
9. ft
10. °F

Goal Practice (page 238)

1. c. 3
2. c. 1,760
3. c. inch
4. a. sq yd
5. d. 7 ft
6. a. 8 ft

Goal Review (page 239)

1. b. 12
2. b. 1,760
3. c. mile
4. b. foot
5. a. inch
6. a. inch
7. a. foot
8. a. 12 in.
9. b. 500 ft
10. d. 6 in.

Goal Practice (page 240)

1. a. 2
2. d. 16
3. a. fluid ounce

4. c. quart
5. a. 1 pt
6. c. 1 fl oz

Goal Review (page 241)

1. d. 16
2. b. 4
3. b. pint
4. a. gallon
5. c. fluid ounces
6. b. pint
7. a. .5 fl oz
8. d. 30 fl oz
9. a. 12 fl oz
10. a. 15 gal

Goal Practice (page 243)

1. c. 2,000
2. b. 16
3. a. ton
4. c. pound
5. c. 1 T
6. a. 2 oz

Goal Review (page 243)

1. b. 16
2. c. 2,000
3. c. ounce
4. a. ton
5. c. ounce
6. a. pound
7. a. 30 oz
8. c. 10 lb
9. a. 5 lb
10. a. .5 lb

Goal Practice (page 245)

1. a. 32°F
2. c. 212°F
3. c. 60°F–70°F
4. a. 32°F
5. d. 400°F
6. b. 68°F

Goal Review (page 245)

1. d. 212°F
2. a. 15°F
3. d. 98.6°F
4. b. 32°F
5. a. 20°F
6. d. 500°F
7. b. 90°F
8. c. 100°F
9. a. 10°F
10. d. 32°F

Auditing Corner (page 246)

1. qt
2. okay
3. okay
4. okay
5. gal
6. fl oz
7. okay
8. okay
9. sq ft
10. okay

Business Corner (page 247)

1. 32,000 lb
2. 500,000 lb
3. 30,000,000 lb
4. 500 lb
5. 130,000 lb

Consumer Corner

Size or Power (page 248)

1. 24,000 Btus
2. 3,264 Btus
3. 768 Btus
4. 30,600 Btus
5. 3,200 Btus

Efficiency (page 250)

1. Bergmart
2. EZ Cool
3. Bergmart
4. EZ Cool
5. Bergmart

• GOAL 12 • RENAMING U.S. CUSTOMARY UNITS

Goal Practice (page 254)

1. 3 ft = 1 yd
2. 1 lb = 16 oz
3. 1 pt = 16 fl oz
4. 1 sq yd = 9 sq ft
5. 1 T = 2,000 lb
6. 4 qt = 1 gal

Goal Review (page 254)

1. 16 oz = 1 lb
2. 1 pt = 16 fl oz
3. 1 qt = 2 pt
4. 1 T = 2,000 lb
5. 16 oz = 1 lb
6. 16 oz = 1 lb

Goal Practice (page 256)

1. a. 1 gal = 4 qt
   b. smaller to larger: divide
   c. 18 qt ÷ 4 qt = $4\frac{1}{2}$ gal
   d. The answer is reasonable because there are fewer larger units.
2. a. 1 qt = 2 pt
   b. smaller to larger: divide
   c. 60 pt ÷ 2 pt = 30 qt
   d. The answer is reasonable because there are fewer larger units.
3. a. 12 in. = 1 ft
   b. larger to smaller: multiply
   c. 12 in. × 33 ft = 396 in.
   d. The answer is reasonable because there are more smaller units.
4. a. 16 oz = 1 lb
   b. larger to smaller: multiply
   c. 16 oz × 5 lb = 80 oz = 80 bowls
   d. The answer is reasonable because there are more smaller units.
5. a. 1 gal = 4 qt
   b. smaller to larger: divide
   c. 10 qt ÷ 4 qt = $2\frac{1}{2}$ gal
   d. The answer is reasonable because there are fewer larger units.
6. a. 1 yd = 3 ft
   b. smaller to larger: divide
   c. 9 ft ÷ 3 ft = 3 yd
   d. The answer is reasonable because there are fewer larger units.

Goal Review (page 257)

1. a. 16 oz = 1 lb
   b. larger to smaller: multiply
   c. 16 oz × 25 lb = 400 oz
   d. The answer is reasonable because there are more smaller units.
2. a. 4 qt = 1 gal
   b. larger to smaller: multiply
   c. 4 qt × 6 gal = 24 qt
   d. The answer is reasonable because there are more smaller units.
3. a. 16 fl oz = 1 pt
   b. larger to smaller: multiply
   c. 16 fl oz × 16 pt = 256 fl oz
   d. The answer is reasonable because there are more smaller units.
4. a. 1 T = 2,000 lb
   b. smaller to larger: divide
   c. 10,000 lb ÷ 2,000 lb = 5 T

d. The answer is reasonable because there are fewer larger units.

5. a. 1 yd = 3 ft
   b. smaller to larger: divide
   c. 25 ft ÷ 3 ft = $8\frac{1}{3}$ yd
   d. The answer is reasonable because there are fewer larger units.
6. a. 16 oz = 1 lb
   b. larger to smaller: multiply
   c. 16 oz × 5 lb = 80 oz = 80 cubes
   d. The answer is reasonable because there are more smaller units.

Goal Practice (page 258)

1. ounces; tons
2. quarts; quarts; fluid ounces
3. square feet; square feet
4. yards; yards
5. 6,480 sq in.
6. 3.125 T
7. 1.894 mi
8. 50 yd
9. .219 T
10. 10 qt

Goal Review (page 259)

1. ounces; tons
2. quarts; quarts; fluid ounces
3. square feet; square feet
4. yards; yards
5. 7.813 gal
6. 77,760 sq in.
7. 3.75 T
8. 2,667 yd
9. 15,840 ft
10. 24

Goal Practice (page 261)

1. 120 grains
2. 1,440 grains
3. 3,840 drams
4. 20.833 pounds
5. 3.571 long tons

Goal Review (page 261)

1. 54 scruples
2. 500 pounds
3. 2,000 pounds
4. 8,960 pounds
5. 10 pennyweights
6. 4 ounces

Auditing Corner (page 262)

1. 6.31 lb
2. 15.6 lb
3. 25 qt
4. 12.5 qt
5. 15 qt
6. okay
7. okay
8. 75 lb
9. okay
10. okay

Business Corner (page 264)

1. a. 216 sq ft
   b. 24 sq yd
2. a. 54 sq ft
   b. 7,776 tiles
3. a. 1,944 sq in.
   b. 1.5 or $1\frac{1}{2}$ sq yd
4. a. 700 sq ft
   b. 77.778 or $77\frac{7}{9}$ sq yd
5. a. 1,200 sq yd
   b. 10,800 sq ft

Consumer Corner (page 265)

1. a. 128 ft
   b. $42\frac{2}{3}$ or 42.667 yd
2. a. 400 ft
   b. $133\frac{1}{3}$ or 133.333 yd
3. a. 280 in.
   b. $7\frac{7}{9}$ or 7.778 yd
4. a. 66 ft
   b. 22 yd

Note: Problem in textbook should read: "a garden 30 feet on each side"

5. a. 120 ft
   b. 40

• WORD PROBLEM STRATEGIES 4 (Page 267)

1. to 10. Answers will vary.

• GOAL 13 • IDENTIFYING SI UNITS

Goal Practice (page 270)

1. SI, or International System of Units
2. a. liter
   b. square meter
   c. gram
   d. degree Celsius
3. a. capacity or volume
   b. area or surface
   c. mass or weight
   d. temperature
4. a. square meter
   b. gram
   c. degrees Celsius

Goal Review (page 271)

1. a. mass or weight
   b. capacity or volume
   c. length or distance
   d. temperature
   e. area or surface
2. a. gram
   b. degrees Celsius
   c. meter
   d. liter
   e. square meter

Goal Practice (page 274)

1. a. one thousand or 1,000
   b. one tenth or $\frac{1}{10}$ or .1
   c. one hundredth, $\frac{1}{100}$, or .01
2. smaller
3. 10 decimeter piece
4. 1 centigram object

Goal Review (page 274)

1. one thousandth, $\frac{1}{1,000}$, or .001 of a gram
2. smaller
3. smaller
4. no
5. hectogram
6. centimeter
7. today
8. deciliter

Goal Practice (page 276)

1. 1
2. mm
3. 600 square meters
4. 7 grams
5. 5 decimeters
6. a. km
   b. kg
   c. ml
   d. °C
   e. dl

Goal Review (page 276)

1. hg
2. °C
3. .33 of a kilogram
4. 50 milliliters

5. 24 degrees Celsius
6. a. dag
   b. kg
   c. ml
   d. °C
   e. dg

Auditing Corner (page 277)

1. 2 kg
2. 4 l
3. 3 l
4. $\frac{1}{2}$ kg
5. 3 kg
6. 1 kg
7. $\frac{1}{2}$ kg
8. 100 g
9. 2 kg
10. 5 m

Business Corner (page 278)

1. b. meters
2. a. degrees Celsius
3. c. kilometers
4. c. m^2
5. b. kilometers
6. c. centimeters
7. b. kilograms
8. d. kilograms
9. a. degrees Celsius
10. c. meters

Consumer Corner (page 280)

1. c. grams
2. b. meters
3. d. degrees Celsius
4. c. centiliters
5. d. square meters
6. a. meters
7. a. meters
8. b. grams
9. c. kilometers
10. b. liters

Goal Practice (page 284)

1. d. kilometer
2. b. cm^2
3. a. meter
4. b. square meter
5. b. decimeter
6. b. millimeter

Goal Review (page 284)

1. b. millimeter
2. b. meter
3. a. m^2
4. d. 2 meters
5. c. m^2
6. a. centimeter

Goal Practice (page 286)

1. a. liter
2. b. milliliter
3. d. liter
4. b. centiliter
5. b. liter
6. d. milliliter

Goal Review (page 287)

1. d. liter
2. c. milliliter
3. a. 20 liters
4. c. centiliter
5. d. liter
6. a. centiliter

Auditing Corner (page 288)

	Qty. Recd.	Qty. Sold	Balance
1.		okay	21 l
2.		10 l	11 l
3.	okay		okay
4.	okay		okay
5.		84 m^2	okay
6.		okay	okay
7.	100 m^2		136 m^2
8.	okay		okay
9.		14 ml	okay
10.		okay	10 ml
11.		okay	3 ml
12.	okay		okay
13.		okay	65 m
14.		6 m	okay
15.		okay	51 m

Business Corner (page 290)

1. b. centimeters
2. b. meters
3. b. centimeters
4. a. m^2
5. a. liters
6. a. liters
7. a. centimeters
8. c. milliliters
9. d. liter
10. a. centimeters

Consumer Corner (page 291)

1. a. 30 centimeters
2. a. 1 milliliter
3. a. 15 centimeters
4. c. 400 liters
5. b. 1 decimeter
6. c. 75 liters
7. b. 200 centimeters
8. a. 355 milliliters
9. a. 25 centimeters
10. a. 200 milliliters

Goal Practice (page 294)

1. b. milligram
2. c. 5 grams
3. b. .2 kilogram
4. d. gram
5. c. kilogram
6. d. kilogram

Goal Review (page 295)

1. b. 3 kilograms
2. d. dekagram
3. c. 50 kilograms
4. c. milligram
5. a. 3 grams
6. c. kilogram

Goal Practice (page 297)

1. c. 100°C
2. b. 0°C
3. a. 0°C
4. b. 35°C
5. d. 200°C
6. a. 0°C

Goal Review (page 297)

1. b. 37°C
2. d. 100°C
3. c. 39°C
4. c. 25°C
5. b. 15°C
6. c. 80°C

Auditing Corner (page 298)

- The ad says it guarantees a weight loss of 5 grams a week and also 5 kilograms a week. It has to be one or the other.
- Also, 25 calories per kilogram multiplied by 70 kilograms would mean 1,750 calories, not 2,570 calories, a day.

Business Corner (page 299)

1. c. kilograms
2. b. kilograms
3. a. kilograms
4. a. grams
5. b. kilograms
6. b. 25
7. d. 37
8. a. 10
9. b. 37
10. d. 90

Consumer Corner (page 300)

1. b. 90 kilograms
2. a. 0°C

3. a. 7 kilograms
4. d. 180°C
5. a. 920 kilograms
6. b. 37°C
7. a. 1 gram
8. a. 15°C
9. b. 1 kilogram
10. 1. 0°C

• GOAL 14 • RENAMING SI UNITS

Goal Practice (page 303)

1. 1 l = 1,000 ml
2. 1 dg = .10 g
3. 1 hm = 100 m
4. 1 km = 1,000 m
5. 1 kg = 1,000 g
6. 1 kl = 1,000 l

Goal Practice (page 306)

1. a. 1 dm = .1 m; 10 dm = 1 m
   b. smaller to larger: divide
   c. 2,000 dm ÷ 10 dm per m = 200 m
   d. The answer is reasonable because there are fewer larger units.
2. a. 1 ml = .001 l; 1,000 ml = 1 l
   b. larger to smaller: multiply
   c. 18 l × 1,000 ml per l = 18,000 ml
   d. The answer is reasonable because there are more smaller units.
3. a. 1 m = 100 cm; 1 m^2 = 10,000 cm^2
   b. smaller to larger: divide
   c. 40,000 cm^2 ÷ 10,000 cm^2 per m^2 = 4 m^2
   d. The answer is reasonable because there are fewer larger units.
4. a. 1 km = 1,000 m
   b. larger to smaller: multiply
   c. 5 km × 1,000 m per km = 5,000 m = 5,000 laces
   d. The answer is reasonable because there are more smaller units.
5. a. 1 ml = .001 l; 1,000 ml = 1 l
   b. smaller to larger: divide
   c. 100 ml × 24 bottles = 2,400 ml
      2,400 ml ÷ 1,000 ml per l = 2.4 l
   d. The answer is reasonable because there are fewer larger units.
6. a. 1 ml = .001 l; 1,000 ml = 1 l
   b. smaller to larger: divide
   c. 500 ml × 10 candles = 5,000 ml
      5,000 ml ÷ 1,000 ml per l = 5 l
   d. The answer is reasonable because there are fewer larger units.

Goal Review (page 307)

1. a. 1 ml = .001 l; 1,000 ml = 1 l
   b. larger to smaller: multiply
   c. 520 l × 1,000 ml per l = 520,000 ml
   d. The answer is reasonable because there are more smaller units.
2. a. 1 kg = 1,000 g
   b. larger to smaller: multiply
   c. 75 kg × 1,000 g per kg = 75,000 g
   d. The answer is reasonable because there are more smaller units.
3. a. 1 cm = .01 m; 1 m = 100 cm
   b. smaller to larger: divide
   c. 350 cm ÷ 100 cm per m = 3.5 m
   d. The answer is reasonable because there are fewer larger units.
4. a. 1 m = 100 cm; 1 m^2 = 10,000 cm^2
   b. larger to smaller: multiply
   c. 8 m^2 × 10,000 cm^2 per m^2 = 80,000 cm^2
   d. The answer is reasonable because there are more smaller units.
5. a. 1 g = 1,000 mg
   b. larger to smaller: multiply
   c. 45 g × 1,000 mg per g = 45,000 mg
   d. The answer is reasonable because there are more smaller units.
6. a. 1 m = 100 cm
   b. larger to smaller: multiply
   c. 300 m × 100 cm per m = 30,000 cm
   d. The answer is reasonable because there are more smaller units.
7. a. 1 km = 1,000 m; 1 km^2 = 1,000,000 m^2
   b. larger to smaller: multiply
   c. 15 km^2 × 1,000,000 m^2 per km^2 = 15,000,000 m^2
   d. The answer is reasonable because there are more smaller units.
8. a. 1 ml = .001 l; 1 l = 1,000 ml
   b. larger to smaller: multiply
   c. 25 l × 1,000 ml per l = 25,000 ml
   d. The answer is reasonable because there are more smaller units.
9. a. 1 kg = 1,000 g
   b. smaller to larger: divide
   c. 5,000 g ÷ 1,000 g per kg = 5 kg
   d. The answer is reasonable because there are fewer larger units.
10. a. 1 kl = 1,000 l
    b. larger to smaller: multiply
    c. 400 kl × 1,000 l per kl = 400,000 l
    d. The answer is reasonable because there are more smaller units.

Goal Practice (page 308)

1. 27,000 mg = 27 g = .027 kg
2. 75,000 ml = 75 l = .075 kl
3. 8,642 cm = 86.42 m = .086 km
4. 50 kg = 50,000 g = 50,000,000 mg
5. 4 kl = 4,000 l = 4,000,000 ml
6. 475 dm = 47.5 m = .048 km

Goal Review (page 309)

1. 5,470 cm = 54.7 m = .055 km
2. 21 kl = 21,000 l = 21,000,000 ml
3. 4,750 mg = 4.75 g = .005 kg
4. 3.5 kg = 3,500 g = 3,500,000 mg
5. .527 km = 527 m = 5,270 dm
6. 1 hl = 100 l = 100,000 ml

Auditing Corner (page 309)

1. a. okay
   b. 300,000
   c. okay
   d. okay
   e. 62,000
2. a. okay
   b. 49,000
   c. 35,000
   d. okay
   e. okay

Business Corner (page 311)

1. 50
2. 16
3. 125
4. 142
5. 233
6. 50
7. 65
8. 124
9. 38
10. 8

Consumer Corner (page 312)

1. a. 5 kg
   b. 14 l
   c. 6 l
   d. 1 kg
   e. 1 kg
2. a. 4 kg
   b. 5 kg
   c. 3 kg
   d. 2 kg
   e. 2 kg
   f. 18 l
   g. 12 l
   h. 3 l

• GOAL 15 • MEASURING WITH U.S. CUSTOMARY AND SI RULERS

Goal Practice (page 316)

1. A
2. $\frac{1}{2}$ unit
3. 1 unit
4. G
5. F
6. $1\frac{1}{2}$ units

Goal Practice (page 318)

1. $\frac{1}{2}$ inch
2. $1\frac{1}{2}$ inches
3. $\frac{1}{2}$ inch
4. 1 inch
5. 1 inch
6. 1 inch

Goal Review (page 319)

1. $\frac{1}{2}$ inch
2. 1 inch
3. 1 inch
4. $\frac{1}{2}$ inch
5. $\frac{1}{2}$ inch
6. 1 inch

Goal Practice (page 320)

1. $\frac{15}{16}$ inch
2. $1\frac{3}{8}$ inches
3. $\frac{1}{2}$ inch
4. $\frac{3}{8}$ inch
5. $1\frac{3}{16}$ inches
6. $\frac{1}{4}$ inch

Goal Review (page 321)

1. $\frac{3}{16}$ inch
2. $1\frac{1}{2}$ inches
3. $2\frac{1}{16}$ inches
4. $\frac{7}{8}$ inch
5. $1\frac{1}{8}$ inches
6. $\frac{3}{16}$ inch

Goal Practice (page 323)

1. 12 millimeters
2. 31 millimeters
3. 4 millimeters
4. 11 millimeters
5. 26 millimeters
6. 36 or 37 millimeters

Goal Review (page 324)

1. 30 millimeters
2. 33 millimeters
3. 27 millimeters
4. 24 millimeters
5. 22 or 23 millimeters
6. 10 millimeters

Auditing Corner (page 324)

1. a. $\frac{11}{16}''$
   b. $\frac{3}{16}''$
   c. $\frac{1}{4}''$
   d. $\frac{1}{4}''$ or $\frac{5}{16}''$
   e. $\frac{5}{8}''$
   f. $\frac{9}{16}''$
   g. $\frac{3}{16}''$
2. a. $\frac{3}{8}''$
   b. okay
   c. okay
   d. $\frac{1}{2}''$
   e. $3\frac{3}{8}''$

Business Corner (page 326)

1. Your drawings should measure:
   a. 4.5 cm × 4.5 cm
   b. 5 cm × 4 cm
   c. 6 cm × 4.5 cm
   d. 3 cm × 3 cm
   e. 2.5 cm × 1.5 cm
   Have someone measure your drawings to check your work.
2. a. 6 meters
   b. 3 meters × 4 meters
   c. 3 meters × 5 meters
   d. 3 meters × 3 meters
   e. 2 meters × 3 meters

Consumer Corner (page 328)

1. a. 25 km
   b. 111 km
   c. 47 km
   d. 70 km
   e. 37 km
2. a. 34 km
   b. 10 km
   c. 7 km
   d. 19 km
   e. 29 km
3. a. 141 mi
   b. 226 mi
   c. 422 mi
   d. 267 mi
   e. 383 mi

• WORD PROBLEM STRATEGIES 5 (Page 334)

1. c
2. b
3. a
4. b
5. a
6. c
7. c
8. a
9. c
10. b

• GOAL 16 • ESTIMATING ANSWERS

Goal Practice (page 338)

1. 9,000
2. 10
3. 300
4. 200
5. 2,400,000
6. 400,000
7. $11,000
8. $8,000
9. $2,000
10. $1,000

Goal Review (page 338)

1. 600
2. 120
3. $3
4. 32,000
5. 14,000
6. 100
7. 900
8. 6,000
9. $600
10. 500

Goal Practice (page 340)

1. a
2. b
3. b
4. b
5. c
6. b
7. c
8. b
9. b
10. b

Goal Review (page 341)

1. a
2. a
3. c
4. b
5. c
6. d
7. b
8. b
9. c
10. b

Goal Practice (page 343)

1. b
2. b
3. a
4. c
5. d
6. c
7. d
8. c
9. c
10. b

Goal Review (page 344)

1. b
2. c
3. d
4. c
5. b
6. c
7. c
8. b
9. c
10. c

Auditing Corner (page 345)

Note: finding these mathematical errors does not tell you which facts are true, only that the arithmetic is wrong. Also, estimates will vary.

1. $156,000,000 − $115,000,000 does *not* equal $21,000,000
2. These foreign sales total about $30,000,000, not $23,000,000.
3. $36,000 × 45 = *at least* $1,500,000
4. $9,000,000 − $6,952,350 = close to $2,000,000; *not* $3,000,000.
5. 27,000 + 23,500 = about 50,000, *not* 60,500.

Business Corner (page 347)

Note: Answers will vary. A suggested range of estimates is given.

1. a. between $3.00 and $3.25
   b. between $3.00 and $3.15
   c. between $1.00 and $1.80
   d. between $1.00 and $1.10
2. between $33 and $37
3. between 89,000,000 and 93,000,000
4. a. between 12,000 and 18,000
   b. between 4,000 and 7,000
   c. between 12,000 and 16,000
   d. between 12,000 and 16,000
5. a. between 690,000 and 730,000
   b. between 1,440,000 and 1,500,000

Consumer Corner (page 348)

Note: Answers will vary. A suggested range of estimates is given.

1. a. between $30 and $50
   b. between $20 and $40
   c. between $20 and $25
   d. between $10 and $20
2. less than $26.00
3. a. between $9,000 and $12,000
   b. between $3,000 and $7,000
   c. between $9,000 and $12,000
4. between $280 and $320

• WORD PROBLEM STRATEGIES 6 (Page 350)

1. 3,000
2. 6%
3. 1,410
4. 12.5%
5. 96
6. 2,210
7. 3,360
8. 200
9. $75
10. $2,500

• GOAL 17 • READING TABLES AND GRAPHS

Goal Practice (page 354)

1. 224
2. 228
3. 234
4. 266
5. 270
6. multiplication table
7. 11
8. 12

Goal Review (page 355)

1. 285
2. 187
3. 168
4. 320
5. 195
6. 12
7. 13

Goal Practice (page 357)

1. $1.36
2. $.22

3. $.64
4. $.72
5. $2.60
6. $2.86

Goal Review (page 359)

1. $2.24
2. $.74
3. $.88
4. $.28
5. $1.66
6. $2.95
7. $.48
8. $1.52
9. $2.80
10. $3.20

Goal Practice (page 361)

1. $21.30
2. $49.60
3. $25.50
4. $36.60
5. $37.00
6. $49.80

Goal Review (page 361)

1. $63.90
2. $34.80
3. $39.20
4. $47.00
5. $34.40
6. $46.80
7. $73.10
8. $52.90
9. $44.40
10. $82.50

Goal Practice (page 364)

1. 9 quarts
2. 5 quarts
3. 3 quarts
4. 7 quarts
5. 5 quarts
6. 8 quarts

Goal Review (page 364)

1. 4 quarts
2. 3 quarts
3. 8 quarts
4. 5 quarts
5. 3 quarts
6. 6 quarts
7. 3 quarts
8. 4 quarts
9. 5 quarts
10. 4 quarts

Auditing Corner (page 365)

1. okay
2. $80.50
3. okay
4. okay
5. $3.70
6. $6.60
7. okay
8. okay
9. $1.70
10. okay

Business Corner (page 368)

1. a. $20.10
   b. $19.56
2. $18.69
3. $25.73
4. $19.24
5. a. $19.83
   b. $25.53
   c. $19.80
   d. $33.27
   e. $13.15
   f. $38.78

Consumer Corner (page 371)

1. $298.60
2. $27.15
3. $1,045.81
4. $298.60
5. $98.75
6. $66.60
7. $132.80

Goal Practice (page 376)

1. a. 20,000
   b. 23,100; 23,200; or 23,300
   c. 24,600; 24,700; or 24,800
2. July
3. January and February
4. 1,000

Goal Review (page 376)

1. a. 23,000
   b. 22,000
   c. 19,500
2. June
3. October to November
4. July to August

Goal Practice (page 378)

1. a. $25 billion
   b. $90 billion
   c. $55 billion
2. food stores
3. package stores
4. $35 billion

Goal Review (page 379)

1. a. $55 billion
   b. $15 billion
   c. $145 billion
2. $90 billion
3. food stores and automotive dealers
4. $240 billion

Goal Practice (page 380)

1. intermediate
2. full size
3. subcompact
4. compact
5. subcompact and compact

Goal Review (page 381)

1. subcompact
2. this year
3. intermediate
4. full size
5. intermediate

Auditing Corner (page 381)

3. Tompkins
8. Dali
9. Fillmore
10. King

Business Corner (page 383)

1. snacks and movies
2. clothing and movies
3. hobbies
4. cosmetics and jewelry
5. male
6. male
7. female
8. snacks
9. male
10. male

Consumer Corner (page 384)

1. Dean Productions
2. go up
3. a. later
   b. beginning
4. $42
5. April 19x1
6. $62
7. a. $66
   b. April 19x2
8. a. April
   b. October
   c. $18

9. a. $48
   b. $52
   c. $56
   d. $58
10. rising

• GOAL 18 • EQUATIONS AND BUSINESS FORMULAS

Goal Practice (page 389)

1. a. not an equation
   b. an equation
   c. an equation
   d. an equation
2. a. 30 × 15 = Distance traveled
   b. 30 + 28 + 30 + 28 = Perimeter
   c. $10,000 ÷ 5 = Individual profit

Goal Review (page 390)

1. a. not an equation
   b. an equation
   c. an equation
   d. an equation
2. a. $54 + $3.78 = Total
   b. $200,000 × 75% = Amount spent
   c. $17.95 − $9.50 = Profit
   d. $79.24 ÷ 4 = Individual share

Goal Practice (page 392)

1. Cost per unit = $65 × 2
2. Price = $78.40 − $7.50
3. Cost = $1,500 − $950
4. Cost = $680 × 4
5. Amount = $\frac{\$120}{6\%}$

Goal Review (page 393)

1. Principal = $\frac{\$37.50}{15\%}$
2. Cost = $600 × 6
3. Amount in stock = 27 + 17
4. Price = $101.25 ÷ .75
5. Pounds per load = 33 + 23

Goal Practice (page 394)

1. Percentage = Base × Rate
   Rate = Percentage ÷ Base
2. Decrease = Base × Rate of decrease
3. Commission = Sales × Rate of commission
   Sales = Commission ÷ Rate of commission
4. Increase = Base × Rate of increase
5. Markup = Cost price × Rate of markup
   Rate of markup = Markup ÷ Cost price
6. Interest = Principal × Rate × Time
   Time = $\frac{\text{Interest}}{\text{Principal} \times \text{Rate}}$

Goal Practice (page 396)

1. 1,100
2. 30%
3. 296
4. 60%
5. 2,050
6. 152
7. 10
8. 8.89%

Goal Review (page 397)

1. 4%
2. 8,050
3. 7.2
4. 500
5. 2%
6. 536.8
7. $45,000
8. 20%
9. 25%
10. 150

Goal Practice (page 398)

1. −$1,320
2. +20%
3. −2.5%
4. $15,000
5. −150
6. +80%
7. +$22,800
8. −$6,750
9. −33.33% or $-33\frac{1}{3}\%$
10. +$25,000

Goal Review (page 399)

1. −7%
2. −$3,000
3. $84,500
4. +13%
5. $1,275
6. +$11.16
7. −$5,120
8. +$32.40
9. +6.11% or $+6\frac{1}{9}$
10. +$87,500

Goal Practice (page 402)

1. $33
2. 20%
3. $357.50
4. $12,000
5. $325
6. 14%
7. $27
8. 6.62%
9. $6
10. 8.22%

Goal Review (page 402)

1. $24
2. $160
3. 15%
4. $6,000
5. 15%
6. $640
7. 44.44%
8. $120
9. 6%
10. $4,375

Goal Practice (page 405)

1. $30
2. 40%
3. $256.47
4. 9%
5. $1,500
6. $248
7. $5.60
8. 66.67%
9. 25%
10. $362,500

Goal Practice (page 405)

1. $78
2. 14%
3. $1.60
4. $938
5. $38.70
6. 11%
7. 150%
8. $60
9. 8.33%
10. $63

Goal Practice (page 406)

1. Interest: $480
   Total due: $3,480
2. Discount: $.98
   Reduced price: $13.01
3. Increase: $2,312.49
   New total: $6,369.49
4. Decrease: $1,730.16
   New amount: $672.84
5. Markup: $112.14
   New selling price: $432.54
6. Interest: $153,900
   Total amount due: $963,900
7. $15.50
8. $2,280

Goal Review (page 407)

1. Increase: $16,740
   New total: $43,740
2. Markup: $83.82
   New selling price: $363.22
3. Discount: $28.34
   Reduced price: $189.66
4. Total amount due: $295,000
5. Discount: $.67
   Reduced price: $6.82
6. Decrease: $.77
   New total: $.72
7. $1,960
8. $1,132.83

Auditing Corner (page 408)

1. Percentage = Base × Rate
   $2 = $.75 × Rate
2. okay
3. okay
4. Percentage = Base × Rate
   150 = Base × 12%
5. okay
6. Percentage = Base × Rate
   $25 = $800 × Rate
7. okay
8. Percentage = Base × Rate
   50 = 1,050 × Rate
9. Percentage = Base × Rate
   $75 = Base × 15%
10. Percentage = Base × Rate
    20 = Base × 12%

Business Corner (page 411)

1. a. Key word: Increase
   b. Formula: Increase = Base × Rate of increase
2. a. Key word: none
   b. Formula: Percentage = Base × Rate
3. a. Key word: Decrease
   b. Formula: Decrease = Base × Rate of decrease
4. a. Key word: Commission
   b. Formula: Commission = Sales × Rate of commission
5. a. Key word: Discount
   b. Formula: Discount = Base price × Rate of discount
6. a. Key word: Markup
   b. Formula: Markup = Cost price × Rate of markup
7. a. Key word: Commission
   b. Formula: Commission = Sales × Rate of commission
8. a. Key word: Interest
   b. Formula: Interest = Principal × Rate × Time
9. a. Key word: none
   b. Formula: Percentage = Base × Rate
10. a. Key word: Discount
    b. Formula: Discount = Base price × Rate of discount

Consumer Corner (page 412)

1. a. Key word: none
   b. Formula: Percentage = Base × Rate
2. a. Key word: Increase
   b. Formula: Increase = Base × Rate of increase
3. a. Key word: Interest
   b. Formula: Interest = Principal × Rate × Time
4. a. Key word: Interest
   b. Formula: Interest = Principal × Rate × Time
5. a. Key word: none
   b. Formula: Percentage = Base × Rate
6. a. Key word: Decrease
   b. Formula: Decrease = Base × Rate of decrease
7. a. Key word: none
   b. Formula: Discount = Base price × Rate of discount
8. a. Key word: none
   b. Formula: Increase = Base × Rate of increase
9. a. Key word: Interest
   b. Formula: Interest = Principal × Rate × Time
10. a. Key word: Interest
    b. Formula: Interest = Principal × Rate × Time

• WORD PROBLEM STRATEGIES 7 (Page 414)

1. Facts A and B are needed.
2. Facts D and E are needed.
3. Missing information.
4. Facts A and B are needed.
5. Facts A, B, and C are needed.
6. Facts A, B, and D are needed.
7. Facts A and B are needed.
8. Facts A, B, and D are needed.
9. Facts A, C, and E are needed.
10. Missing information.

• GOAL 19 • SOLVING WORD PROBLEMS

Goal Practice (page 418)

1. a. Known: $149 paycheck; $50 spent
   Unknown: amount left
   b. (2) subtracting
2. a. Known: $500 down payment; $1,000 total price
   Unknown: what percent of total price the down payment is
   b. (6) percent
3. a. Known: July expenses—$3,542; $4,039; $7,821; $1,893; $1,972
   Unknown: Total expenses for July
   b. (1) adding
4. a. Known: $200 weekly salary
   Unknown: yearly salary
   b. (3) multiplying
5. a. Known: $4.50 hourly rate; $1\frac{1}{2}$ times hourly rate for overtime
   Unknown: overtime rate
   b. (3) multiplying

Goal Review (page 419)

1. a. Known: $50 total tips; 4 attendants
   Unknown: each attendant's share of tips
   b. (4) dividing
2. a. Known: $\frac{9}{10}$
   Unknown: number expressed as a decimal
   b. (5) renaming
3. a. Known: $.45 a yard for wrapping paper; 10 yards wanted
   Unknown: cost of 10 yards
   b. (3) multiplying
4. a. Known: $.15 a check; 9 checks written
   Unknown: fee for 9 checks
   b. (3) multiplying
5. a. Known: spent $115,070; budgeted $104,500
   Unknown: amount over budget
   b. (2) subtracting
6. a. Known: 5,000 lamps cost $10,000 to produce
   Unknown: cost of one lamp
   b. (4) dividing
7. a. Known: $15,000 yearly salary
   Unknown: weekly salary
   b. (4) dividing
8. a. Known: $2.52 per gallon
   Unknown: price per quart
   b. (4) dividing
9. a. Known: $10, $15.50, $3.49, $.59 petty cash paid out
   Unknown: total petty cash paid out
   b. (1) adding

10. a. Known: $20 per 1,000 copies
Unknown: rate per copy
b. (4) dividing

Goal Practice (page 420)

1. b
2. a
3. c
4. b
5. a
6. b

Goal Review (page 422)

1. b, c
2. a, b, c
3. a, b
4. a, b, d
5. a, b, c, d, e
6. a, b, d
7. a, b, c, e, f
8. b, c, f

Goal Practice (page 424)

1. $39,317
2. $26.75
3. $4.14
4. $581.25
5. 25%
6. $320

Goal Review (page 425)

1. $12.90
2. $839.42
3. $.10
4. $265.20
5. $410.20
6. $750
7. 926
8. $6,500
9. $.61
10. $29.98

Goal Practice (page 426)

1. $21
2. 2 gallons
3. $355.25
4. $4.80
5. $3.10
6. $42

Goal Review (page 427)

1. $177.33
2. $20.80
3. $5
4. $387.50
5. $190,000
6. $279.50

Auditing Corner (page 428)

1. A
2. B
3. A
4. A
5. B
6. A
7. B
8. B
9. A
10. B

Business Corner (page 432)

Part I

1. b
2. c
3. b
4. b
5. c

Part II

6. b
7. a
8. b
9. b
10. a

Part III

1. April 5
2. 160
3. $51.25
4. $400
5. $275
6. more facts needed
7. $15,800
8. more facts needed
9. more facts needed
10. 4,000

Consumer Corner (page 434)

1. a, b, e
2. b, d
3. b, d
4. b, c, d
5. a, b, c, d
6. a, b, c; $38.40
7. b, c, d; credit card $3.80
8. b, c; $8.25
9. a, b, c; $24.75
10. a, b, c; $217.25

• GOAL 20 • RENAMING COMMON FRACTIONS IN SIMPLEST TERMS

Goal Practice (page 439)

1. a. $\frac{4}{16}$
b. $\frac{3}{8}$
c. $\frac{3}{6}$
2. a. $\frac{4}{6}$
b. $\frac{4}{5}$
3. $\frac{150,000}{250,000}$
4. $\frac{12,500}{37,500}$
5. $\frac{330,000}{3,300,000}$
6. $\frac{77,600,000}{77,800,000}$
7. $\frac{19,300,000}{76,000,000}$

Goal Practice (page 442)

1. .3333
2. .8
3. .75
4. .6
5. .1667
6. .625
7. $\frac{3}{12}$ or .25
8. auditing $\frac{18}{24}$ or .75

Goal Practice (page 444)

1. a. $\frac{10}{15}$
b. $\frac{15}{40}$
c. $\frac{10}{25}$
2. a. $\frac{3}{4}$
b. $\frac{4}{5}$
c. $\frac{1}{3}$
3. $\frac{2}{3}$
4. $\frac{4}{5}$
5. $\frac{4}{7}$

Goal Review (page 445)

1. a. $\frac{7}{11}$
b. $\frac{3}{12}$
c. $\frac{3}{14}$
2. a. $\frac{70}{130}$
b. $\frac{30}{40}$
c. $\frac{50}{90}$
3. $\frac{5}{6}$
4. $\frac{5}{6}$
5. $\frac{5}{9}$

Goal Practice (page 446)

(Answers will vary.)

1. $\frac{120}{480} = \frac{12}{48} = \frac{1}{4}$
2. $\frac{13}{65} = \frac{1}{5}$
3. $\frac{21}{49} = \frac{3}{7}$
4. $\frac{800}{7{,}200} = \frac{8}{72} = \frac{1}{9}$
5. $\frac{8}{28} = \frac{2}{7}$
6. $\frac{50}{150} = \frac{1}{3}$
7. $\frac{35}{42} = \frac{5}{6}$
8. $\frac{33}{88} = \frac{3}{8}$

Goal Review (page 447)

(Answers will vary.)

1. $\frac{150}{900} = \frac{15}{90} = \frac{3}{18} = \frac{1}{6}$
2. $\frac{45}{81} = \frac{5}{9}$
3. $\frac{17}{51} = \frac{1}{3}$
4. $\frac{22}{55} = \frac{2}{5}$
5. $\frac{14}{49} = \frac{2}{7}$
6. $\frac{35}{50} = \frac{7}{10}$
7. $\frac{8}{36} = \frac{2}{9}$
8. $\frac{1{,}300}{7{,}800} = \frac{13}{78} = \frac{1}{6}$
9. $\frac{3{,}200}{8{,}000} = \frac{32}{80} = \frac{8}{20} = \frac{2}{5}$
10. $\frac{1{,}600}{8{,}000} = \frac{16}{80} = \frac{4}{20} = \frac{1}{5}$

Goal Practice (page 449)

1. $\frac{1}{2}$
2. $\frac{3}{7}$
3. $\frac{3}{8}$
4. $\frac{4}{5}$
5. $\frac{1}{6}$
6. $\frac{2}{3}$
7. $\frac{7}{10}$
8. $\frac{7}{8}$

Goal Review (page 449)

1. $\frac{3}{4}$
2. $\frac{1}{8}$
3. $\frac{1}{4}$
4. $\frac{1}{5}$
5. $\frac{1}{5}$
6. $\frac{11}{12}$
7. $\frac{25}{27}$
8. $\frac{3}{5}$
9. $\frac{5}{8}$
10. $\frac{9}{10}$

Auditing Corner (page 450)

1. okay
2. okay
3. okay
4. $\frac{15}{38}$
5. okay
6. okay
7. okay
8. okay
9. okay
10. $\frac{4}{95}$

Business Corner (page 451)

1. a. $\frac{320}{925} = \frac{64}{185}$
   b. $\frac{210}{925} = \frac{42}{185}$
   c. $\frac{95}{925} = \frac{19}{185}$
   d. $\frac{540}{925} = \frac{108}{185}$
   e. $\frac{205}{925} = \frac{41}{185}$
2. a. $\frac{42}{500} = \frac{21}{250}$
   b. $\frac{370}{500} = \frac{37}{50}$
   c. $\frac{200}{500} = \frac{2}{5}$
   d. $\frac{400}{500} = \frac{4}{5}$
   e. $\frac{155}{500} = \frac{31}{100}$

Consumer Corner (page 452)

1. a. $\frac{11}{80}$
   b. $\frac{10}{118} = \frac{5}{59}$
   c. $\frac{16}{119}$
   d. $\frac{5}{140} = \frac{1}{28}$
   e. $\frac{12}{50} = \frac{6}{25}$
2. a. $\frac{50}{270} = \frac{5}{27}$
   b. $\frac{13}{112}$
   c. $\frac{22}{150} = \frac{11}{75}$
   d. $\frac{18}{28} = \frac{9}{14}$
   e. $\frac{40}{55} = \frac{8}{11}$

• WORD PROBLEM STRATEGIES 8 (Page 454)

1. $49,399
2. $76.75
3. $560.00
4. $30.80
5. $1,206.68
6. $133.88
7. $485.50
8. $585.00
9. $9,300
10. $239.90

• GOAL 21 • RENAMING A MIXED NUMBER AS A COMMON FRACTION

Goal Practice (page 459)

1. $\frac{75}{3}$
2. $\frac{48}{8}$
3. $\frac{190}{10}$
4. $\frac{65}{5}$
5. $\frac{27}{3}$
6. $\frac{112}{4}$
7. 32 quarters
8. 60 twelfths

Goal Review (page 459)

1. $\frac{74}{2}$
2. $\frac{60}{5}$
3. $\frac{352}{4}$
4. $\frac{310}{10}$
5. $\frac{176}{8}$
6. $\frac{170}{10}$
7. 28 quarters
8. 64 eighths
9. 54 sixths
10. 15 fifths

Goal Practice (page 461)

1. $\frac{15}{4}$
2. $\frac{14}{3}$
3. $\frac{13}{3}$
4. $\frac{19}{8}$
5. $\frac{11}{2}$

6. $\frac{9}{2}$
7. 219 eighths
8. 225 tenths

Goal Review (page 461)

1. $\frac{21}{4}$
2. $\frac{67}{10}$
3. $\frac{47}{5}$
4. $\frac{17}{2}$
5. $\frac{23}{3}$
6. $\frac{15}{4}$
7. 115 quarters
8. 9 halves
9. 95 quarters
10. 37 tenths

Goal Practice (page 463)

1. $17\frac{1}{2}$
2. $3\frac{4}{5}$
3. $13\frac{1}{3}$
4. $5\frac{7}{10}$
5. $4\frac{3}{4}$
6. $2\frac{7}{8}$
7. $18\frac{1}{2}$
8. $3\frac{1}{4}$

Goal Review (page 463)

1. $18\frac{2}{3}$
2. $1\frac{6}{8} = 1\frac{3}{4}$
3. $11\frac{1}{4}$
4. $1\frac{8}{10} = 1\frac{4}{5}$
5. $16\frac{3}{5}$
6. $23\frac{2}{3}$
7. $7\frac{2}{5}$
8. $9\frac{7}{10}$
9. $11\frac{1}{4}$
10. $9\frac{3}{8}$

Auditing Corner (page 464)

1. a. $\frac{47}{4}$
   b. okay
   c. okay
   d. $\frac{23}{4}$
   e. okay
2. a. okay
   b. $\frac{101}{8}$
   c. okay
   d. $\frac{59}{8}$
   e. okay

Business Corner (page 465)

1. a. $\frac{3}{2}$
   b. $\frac{11}{4}$
   c. $\frac{14}{3}$
   d. $\frac{5}{2}$
   e. $\frac{17}{4}$
2. a. $\frac{21}{2}$
   b. $\frac{35}{4}$
   c. $\frac{17}{4}$
   d. $\frac{15}{8}$
   e. $\frac{7}{3}$

Consumer Corner (page 466)

1. a. $\frac{93}{10}$
   b. $\frac{29}{5}$
   c. $\frac{22}{3}$
   d. $\frac{103}{8}$
   e. $\frac{47}{4}$
2. a. $\frac{46}{3}$
   b. $\frac{59}{4}$
   c. $\frac{65}{4}$
   d. $\frac{38}{3}$
   e. $\frac{81}{4}$

• GOAL 22 • MULTIPLYING COMMON FRACTIONS AND MIXED NUMBERS

Goal Practice (page 470)

1. $\frac{4}{6} = \frac{2}{3}$
2. $3\frac{1}{2}$
3. $\frac{3}{20}$
4. 6
5. $\frac{10}{24} = \frac{5}{12}$
6. $1\frac{1}{8}$
7. 8
8. $\frac{1}{32}$

Goal Review (page 470)

1. $\frac{8}{15}$
2. $8\frac{1}{2}$
3. 12
4. $\frac{15}{24} = \frac{5}{8}$
5. $5\frac{6}{10} = 5\frac{3}{5}$
6. $8\frac{1}{4}$
7. 1
8. \$7
9. \$3
10. $\frac{1}{20}$

Goal Practice (page 472)

1. $33\frac{1}{2}$
2. $29\frac{1}{5}$
3. $191\frac{2}{8} = 191\frac{1}{4}$
4. $24\frac{3}{8}$
5. 55
6. 44
7. \$1,162.50 or \$1,162$\frac{1}{2}$
8. \$4,472

Goal Review (page 472)

1. $2\frac{6}{24} = 2\frac{1}{4}$
2. $1\frac{13}{20}$
3. $26\frac{2}{3}$
4. 3
5. $39\frac{17}{32}$
6. $12\frac{2}{24} = 12\frac{1}{12}$
7. \$93
8. $9\frac{3}{8}$
9. \$3,110
10. \$1,787.50 or \$1,787$\frac{1}{2}$

Goal Practice (page 474)

1. \$24.70
2. \$16.13
3. \$49.60
4. \$43.58
5. \$39.38
6. \$26.13
7. \$100.40
8. \$87.50

Goal Review (page 474)

1. \$25.80
2. \$69.44

3. $28.33
4. $11.25
5. $13.17
6. $10.95
7. $468.94
8. $2,800
9. $3.19
10. $6,902

Auditing Corner (page 475)

1. a. okay
   b. $36.44
   c. $37.13
2. a. $29.25
   b. $30.38
   c. okay
3. a. $9.15
   b. $7.35
   c. $7.58
   d. $6.75

Business Corner

Stock Purchases (page 477)

1. $12,477.50
2. $4,912.50
3. $4,468.75
4. $3,031.88

Profit and Loss (page 478)

1. a. Blaze $426,140; Gnor $106,535; Sture $106,535
   b. $25,361.17
   c. Gnor $81,250.83; Sture $16,250.17
   d. $16,000,000
   e. $20,000,000
   f. $4,000,000

Consumer Corner

Reduced Prices (page 480)

1. $417.33; $834.67
2. $38; $57
3. $89.81; $29.94
4. $24; $72
5. $24; $72

Circumference and Area (page 481)

1. 44 ft
2. a. $7\frac{1}{14}$ m²
   b. $14\frac{9}{14}$ m²
3. a. $12\frac{4}{7}$ m
   b. 11 m
   c. $6\frac{2}{7}$ m
   d. $9\frac{3}{7}$ m
4. a. $78\frac{4}{7}$ sq ft
   b. 1,386 sq ft
   c. $1{,}307\frac{3}{7}$ sq ft
   d. $31\frac{3}{7}$ ft

• WORD PROBLEM STRATEGIES 9 (Page 484)

1. C
2. B
3. A
4. B
5. A
6. C
7. A
8. C
9. C
10. A

• GOAL 23 • DIVIDING WITH FRACTIONS

Goal Practice (page 490)

1. 30
2. 16
3. $1\frac{3}{5}$
4. $33\frac{1}{3}$
5. $21\frac{5}{15} = 21\frac{1}{3}$
6. 300 miles
7. 20
8. $8,000

Goal Review (page 490)

1. $\frac{15}{16}$
2. 15,000
3. $13\frac{1}{3}$
4. $1\frac{4}{10} = 1\frac{2}{5}$
5. $\frac{10}{12} = \frac{5}{6}$
6. $6,000
7. $10\frac{2}{3}$
8. $2,250
9. $18.75 or $\$18\frac{3}{4}$
10. $162.40 or $\$162\frac{2}{5}$

Goal Practice (page 491)

1. $4\frac{4}{5}$
2. 5
3. $10
4. $33\frac{1}{3}$
5. $5\frac{5}{15} = 5\frac{1}{3}$
6. 40 hours
7. 30
8. 40%
9. 20
10. 40

Goal Review (page 492)

1. $3\frac{11}{23}$
2. $4\frac{74}{84} = 4\frac{37}{42}$
3. $38.25 or $\$38\frac{1}{4}$
4. 330
5. $53\frac{13}{39} = 53\frac{1}{3}$
6. 2,000
7. 120
8. 6
9. 240
10. 10 hours

Goal Practice (page 493)

1. a. divide
   b. $320 \div \frac{5}{8}$
   c. about 500
2. a. multiply
   b. $\$5{,}363{,}540 \times \frac{3}{8}$
   c. about $2 million
3. a. multiply
   b. $5{,}000 \times \frac{3}{10}$
   c. about 1,500
4. a. multiply
   b. $\$300{,}000 \times \frac{4}{5}$
   c. about $240,000
5. a. divide
   b. $\$150{,}000 \div \frac{3}{8}$
   c. about $400,000

Goal Review (page 494)

1. a. divide
   b. $200 \div \frac{4}{5}$
   c. about 250
2. a. divide
   b. $70 \div \frac{7}{8}$
   c. about 80
3. a. multiply
   b. $20 \times \frac{3}{8}$
   c. about 8
4. a. divide
   b. $140 \div \frac{7}{10}$
   c. about 200
5. a. multiply
   b. $\$240 \times \frac{7}{8}$
   c. about $210

Auditing Corner (page 496)

1. a. 32
   b. 49
   c. okay
   d. okay
   e. okay
2. a. okay
   b. 51
   c. okay
   d. okay
   e. 41

Business Corner (page 497)

1. by attendants 25, by customers 22
2. by attendants 127, by customers 112
3. 48
4. a. 67
   b. 51
   c. 46
   d. 24

Consumer Corner (page 498)

1. $33\frac{1}{3}$
2. $12\frac{1}{2}$
3. 10
4. 68
5. 34
6. 16
7. $56\frac{1}{4}$
8. $46\frac{2}{3}$
9. $41\frac{2}{3}$
10. 325

• GOAL 24 • ADDING FRACTIONS

Goal Practice (page 502)

1. $1\frac{2}{8} = 1\frac{1}{4}$
2. $1\frac{5}{10} = 1\frac{1}{2}$
3. $3\frac{1}{5}$
4. $1\frac{5}{8}$
5. $1\frac{9}{10}$
6. $1\frac{3}{4}$
7. $\frac{5}{8}$
8. $\frac{5}{10} = \frac{1}{2}$

Goal Review (page 502)

1. $4\frac{1}{2}$
2. $2\frac{1}{3}$
3. $2\frac{4}{8} = 2\frac{1}{2}$
4. 3
5. $2\frac{3}{10}$
6. $2\frac{1}{4}$ hours
7. 1 million
8. $\frac{3}{10}$

Goal Practice (page 505)

1. $\frac{23}{24}$
2. $\frac{17}{20}$
3. $\frac{14}{16} = \frac{7}{8}$
4. $1\frac{5}{12}$
5. $1\frac{2}{15}$
6. $\frac{14}{16} = \frac{7}{8}$
7. $\frac{11}{15}$
8. $\frac{20}{32} = \frac{5}{8}$

Goal Review (page 505)

1. $\frac{13}{20}$
2. $1\frac{2}{16} = 1\frac{1}{8}$
3. $1\frac{11}{30}$
4. $1\frac{5}{12}$
5. $\frac{31}{40}$
6. yes; $\frac{3}{8} + \frac{2}{8} = \frac{5}{8}$
7. $\frac{6}{8} = \frac{3}{4}$
8. $1\frac{25}{50} = 1\frac{1}{2}$
9. $1\frac{20}{32} = 1\frac{5}{8}$
10. $\frac{20}{32} = \frac{5}{8}$

Goal Practice (page 506)

1. $1\frac{1}{6}$
2. $\frac{7}{8}$
3. $1\frac{5}{8}$
4. $2\frac{1}{10}$
5. $2\frac{1}{4}$
6. $\frac{8}{10} = \frac{4}{5}$
7. 1
8. $\frac{7}{8}$

Goal Review (page 507)

1. $1\frac{5}{10} = 1\frac{1}{2}$
2. $2\frac{3}{8}$
3. $1\frac{4}{10} = 1\frac{2}{5}$
4. $1\frac{1}{4}$
5. $1\frac{1}{8}$
6. $\frac{7}{8}$ lb
7. $1\frac{1}{4}$
8. $1\frac{3}{8}$
9. $1\frac{2}{10} = 1\frac{1}{5}$
10. $1\frac{1}{8}$

Goal Practice (page 508)

1. 17
2. $12\frac{3}{8}$
3. $4\frac{9}{10}$
4. $22\frac{1}{6}$
5. $13\frac{1}{30}$
6. $70\frac{1}{4}$
7. $53\frac{7}{12}$
8. $12\frac{1}{4}$

Goal Review (page 508)

1. $21\frac{5}{8}$
2. $10\frac{1}{6}$
3. $9\frac{6}{8} = 9\frac{3}{4}$
4. $15\frac{1}{6}$
5. $0\frac{7}{10}$
6. $18\frac{3}{4}$
7. $5\frac{7}{12}$
8. $12\frac{7}{8}$
9. $10\frac{1}{4}$
10. $28\frac{5}{12}$

Goal Practice (page 510)

1. $25\frac{8}{10} = 25\frac{4}{5}$
2. $25\frac{3}{8}$
3. $27\frac{22}{40} = 27\frac{11}{20}$
4. $34\frac{1}{8}$
5. $19\frac{5}{6}$
6. $76\frac{3}{10}$
7. $\$20\frac{3}{4}$
8. $441\frac{11}{20}$

Goal Review (page 510)

1. $3\frac{53}{120}$
2. $5\frac{3}{10}$
3. $3\frac{79}{120}$

4. $8\frac{43}{120}$
5. $5\frac{23}{120}$
6. $2\frac{21}{40}$
7. $6\frac{7}{8}$
8. $39\frac{1}{4}$
9. $12\frac{37}{120}$
10. $3\frac{1}{48}$

Auditing Corner (page 511)

1. a. okay
   b. $11\frac{3}{10}$
2. a. okay
   b. $9\frac{23}{60}$ *or* okay
3. a. okay
   b. $5\frac{17}{20}$
4. a. $8\frac{13}{15}$
   b. $6\frac{19}{120}$
5. a. okay
   b. $13\frac{3}{20}$
   c. $9\frac{23}{40}$

Business Corner (page 513)

1. White: 39 hr
2. Schacher: 37 hr
3. Harris: 42 hr
4. Duane: $41\frac{1}{4}$ hr
5. Slackmeyer: 40 hr
6. Fabbricante: $42\frac{3}{4}$ hr

Consumer Corner (page 516)

A. $73\frac{11}{40}$
B. $27\frac{19}{30}$
C. $50\frac{113}{120}$
D. $109\frac{61}{120}$
E. $36\frac{13}{24}$

• WORD PROBLEM STRATEGIES 10 (Page 517)

1. $40.50
2. $629.30
3. $4.72
4. $42.00
5. $30.00
6. $694.30
7. $169,744
8. $1,624
9. 18.5
10. $177.33

• GOAL 25 • SUBTRACTING FRACTIONS

Goal Practice (page 521)

1. $\frac{5}{12}$
2. $\frac{1}{8}$
3. $\frac{7}{30}$
4. $\frac{3}{10}$
5. $\frac{7}{24}$
6. $\frac{1}{10}$
7. $\frac{1}{20}$ pound
8. $\frac{6}{16} = \frac{3}{8}$ pound

Goal Review (page 522)

1. $\frac{2}{8} = \frac{1}{4}$
2. $\frac{5}{10} = \frac{1}{2}$
3. $\frac{7}{40}$
4. $\frac{1}{10}$
5. $\frac{1}{24}$
6. $\frac{2}{15}$
7. $\frac{1}{10}$
8. $\frac{5}{24}$ second

Goal Practice (page 523)

1. $2\frac{1}{8}$
2. $2\frac{1}{4}$
3. $1\frac{3}{5}$
4. $10\frac{1}{3}$
5. $14\frac{1}{2}$
6. $32\frac{7}{10}$
7. $2\frac{3}{4}$ pages
8. $\frac{2}{3}$ case

Goal Review (page 523)

1. $3\frac{3}{5}$
2. $15\frac{3}{10}$
3. $3\frac{1}{4}$
4. $17\frac{1}{8}$
5. $38\frac{2}{3}$
6. $8\frac{1}{2}$
7. $\frac{7}{10}$ pound
8. $\$1\frac{1}{8}$
9. $3\frac{3}{4}$ minutes
10. $2\frac{1}{4}$ cents or $\$.02\frac{1}{4}$

Goal Practice (page 525)

1. $6\frac{3}{8}$
2. $9\frac{4}{10} = 9\frac{2}{5}$
3. $4\frac{7}{10}$
4. $\frac{7}{10}$
5. $9\frac{7}{12}$
6. $2\frac{5}{6}$
7. $18\frac{3}{4}$
8. $4\frac{1}{3}$ hours

Goal Review (page 525)

1. $10\frac{7}{24}$
2. $2\frac{5}{8}$
3. $9\frac{4}{5}$
4. $5\frac{11}{40}$
5. $1\frac{2}{3}$
6. $16\frac{7}{10}$
7. $21\frac{5}{12}$ yards
8. $3\frac{1}{6}\%$
9. $2\frac{2}{40} = 2\frac{1}{20}$ mpg
10. $28\frac{2}{10} = 28\frac{1}{5}$ lb

Auditing Corner (page 526)

1. okay
2. okay
3. $34\frac{5}{12}$
4. $31\frac{2}{12} = 31\frac{1}{6}$
5. $41\frac{3}{4}$
6. $39\frac{5}{12}$
7. $34\frac{11}{12}$
8. $20\frac{7}{12}$
9. $16\frac{1}{12}$
10. okay

Business Corner (page 527)

1. a. $2\frac{5}{8}$ gal
   b. $18\frac{8}{10} = 18\frac{4}{5}$ lb
   c. $6\frac{11}{12}$ dozen
   d. $12\frac{11}{12}$ lb
   e. $5\frac{3}{20}$ lb
   f. $8\frac{8}{10}$ lb $= 8\frac{4}{5}$ lb
   g. $27\frac{7}{8}$ lb
2. a. $\frac{5}{8}$ lb
   b. $\frac{1}{4}$ lb

c. $\frac{3}{8}$ lb

d. $\frac{2}{4} = \frac{1}{2}$ lb

Consumer Corner (page 528)

1. a. $-1\frac{1}{8}$

b. $+15\frac{6}{8} = +15\frac{3}{4}$

c. $+\frac{2}{4} = +\frac{1}{2}$

d. $-6\frac{6}{8} = -6\frac{3}{4}$

2. a. $+2\frac{2}{8} = +2\frac{1}{4}$

b. $+13\frac{2}{4} = +13\frac{1}{2}$; sell

c. $-1\frac{4}{8} = -1\frac{1}{2}$

d. $+6\frac{2}{4} = +6\frac{1}{2}$; sell

ANSWER KEY TO LEARNING GUIDE SKILL BUILDERS

• GOAL 1

Skill Builder (page 3)

Ex. Y
1. O
2. U
3. R
4. F
5. I
6. N
7. G
8. E
9. R
10. S

• GOAL 2

Skill Builder (page 7)

1. D
2. O
3. N
4. T
5. B
6. I
7. T
8. E
9. A
10. N
11. Y
12. I
13. N
14. S
15. E
16. C
17. T
18. S

• GOAL 3

Skill Builder for Section 1 (page 11)

1.

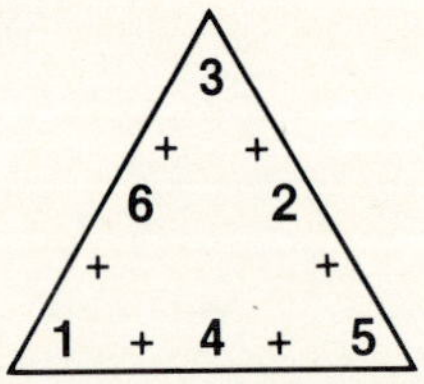

2. a.

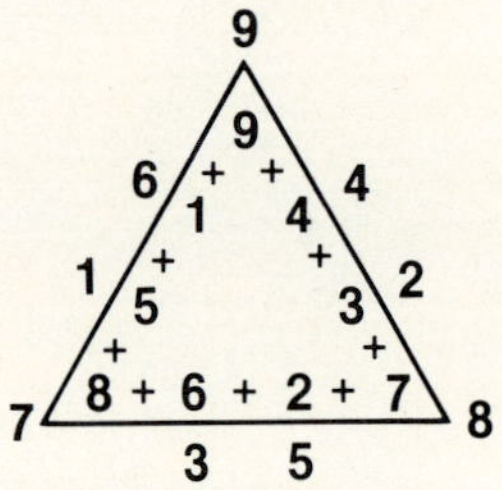

b.

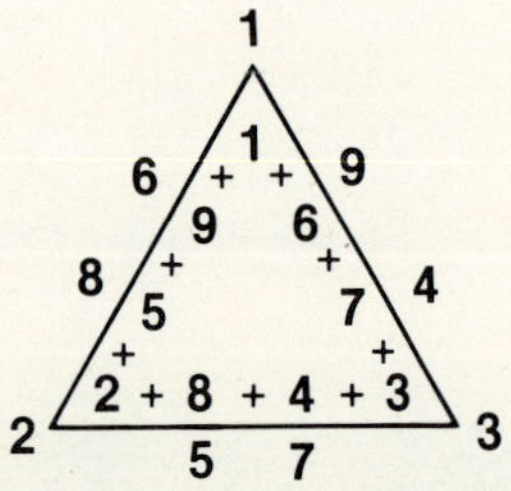

Skill Builder for Section 2 (page 13)

1. Magic sum = 1844

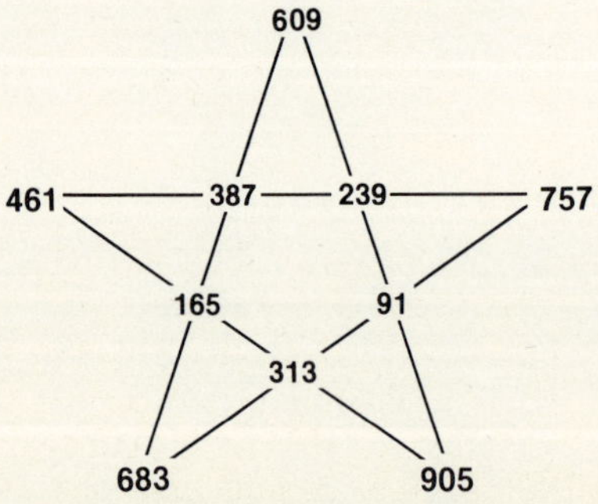

2. Magic sum = 1876

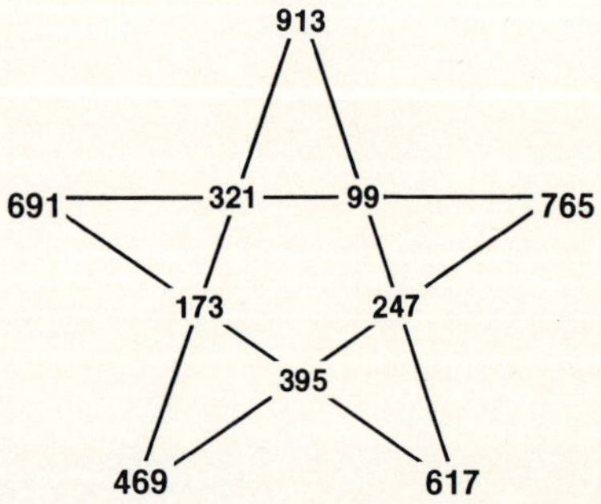

• GOAL 4

Skill Builder for Section 1 (page 17)

Ex. H
1. U
2. M
3. A
4. N
5. R
6. I
7. G
8. H
9. T
10. S

Skill Builder for Section 2 (page 19)

1. a. yes; 297
   b. yes; 297
   c. yes; 297
   d. yes; 297
   e. yes; 297
2. a. yes; 693
   b. yes; 99
   c. yes; 198
   d. yes; 396
   e. yes; 693

• GOAL 5 • Skill Builder for Section 1 (page 27)

×	2	8	4	7	6	9	5	3
8	16	64	32	56	48	72	40	24
3	6	24	12	21	18	27	15	9
6	12	48	24	42	36	54	30	18
4	8	32	16	28	24	36	20	12
2	4	16	8	14	12	18	10	6
9	18	72	36	63	54	81	45	27
7	14	56	28	49	42	63	35	21
5	10	40	20	35	30	45	25	15

Skill Builder for Section 2 (page 29)

Trick 1. Answers will vary.
Trick 2. Answers will vary.

• GOAL 6 • Skill Builder for Section 1 (page 33)

BOARD 1

7 R1 64 ÷ 9	9 R2 74 ÷ 8	9 R2 65 ÷ 7
9 R5 59 ÷ 6	9 R8 89 ÷ 9	5 R4 49 ÷ 9
7 R4 60 ÷ 8	10 R1 71 ÷ 7	6 R4 40 ÷ 6

BOARD 2

3 R4 19 ÷ 5	8 R3 67 ÷ 8	7 R2 51 ÷ 7
6 R1 19 ÷ 3	5 R3 23 ÷ 4	5 R2 47 ÷ 9
9 R1 19 ÷ 2	7 R2 37 ÷ 5	6 R5 41 ÷ 6

BOARD 3

6 R1 13 ÷ 2	10 R5 95 ÷ 9	6 R6 48 ÷ 7
9 R3 57 ÷ 6	8 R1 25 ÷ 3	8 R2 42 ÷ 5
10 R7 87 ÷ 8	8 R3 35 ÷ 4	9 R3 84 ÷ 9

BOARD 4

3 R3 27 ÷ 8	5 R1 36 ÷ 7	5 R3 28 ÷ 5
6 R3 39 ÷ 6	5 R2 17 ÷ 3	4 R6 38 ÷ 8
6 R1 55 ÷ 9	4 R1 13 ÷ 3	8 R1 17 ÷ 2

BOARD 5

4 R7 39 ÷ 8	3 R2 29 ÷ 9	2 R5 19 ÷ 7
5 R3 33 ÷ 6	4 R3 23 ÷ 5	3 R1 13 ÷ 4
5 R1 16 ÷ 3	6 R2 26 ÷ 4	7 R1 36 ÷ 5

BOARD 6

7 R1 15 ÷ 2	7 R2 23 ÷ 3	7 R3 31 ÷ 4
7 R4 39 ÷ 5	7 R5 47 ÷ 6	7 R6 55 ÷ 7
7 R7 63 ÷ 8	7 R8 71 ÷ 9	9 R7 79 ÷ 8

Skill Builder for Section 2 (page 35)

Across
1. 1913
3. 1947
6. 2209
7. 1633
9. 1969
10. 1874
11. 476
12. 2556

Down
2. 1732
4. 4536
5. 1929
6. 2664
7. 1898
8. 1876
9. 1895

• GOAL 7

Skill Builder for Section 1 (page 39)

1. Magic Sum = 9.50

0.5	4.0	3.75	1.25
3.25	1.75	2.0	2.5
2.25	2.75	3.0	1.5
3.5	1.0	0.75	4.25

2. Magic Sum = 16.50

8.25	3.75	3.50	1.00
3.00	1.50	1.75	10.25
2.00	10.50	2.75	1.25
3.25	0.75	8.50	4.00

Skill Builder for Section 2 (page 41)

1. 1.9
×4
7.6
+1.0
8.6
+1.9
10.5
÷5
2.1
−1.9
.2

2. 3.3
×4
13.2
+1.0
14.2
+3.3
17.5
÷5
3.5
−3.3
.2

3. 2.6
×4
10.4
+1.0
11.4
+2.6
14.0
÷5
2.8
−2.6
.2

4. 7.5
×4
30.0
+1.0
31.0
+7.5
38.5
÷5
7.7
−7.5
.2

5. 8.3
×4
33.2
+1.0
34.2
+8.3
42.5
÷5
8.5
−8.3
.2

6. .30
×3
.90
+.05
.95
×5
4.75
−.25
4.50
÷15
.30

7. .42
×3
1.26
+.05
1.31
×5
6.55
−.25
6.30
÷15
.42

8. .64
×3
1.92
+.05
1.97
×5
9.85
−.25
9.60
÷15
.64

9. .28
×3
.84
+.05
.89
×5
4.45
−.25
4.20
÷15
.28

10. .03
×3
.09
+.05
.14
×5
.70
−.25
.45
÷15
.03

• GOAL 8

Skill Builder (page 45)

1. H
2. A
3. V
4. E
5. A
6. P
7. O
8. I
9. N
10. T

• GOAL 9

Skill Builder (page 55)

1. B
2. E
3. C
4. A
5. U
6. S
7. E
8. A
9. W
10. E
11. L
12. L
13. I
14. S
15. A
16. H
17. O
18. L
19. E

• GOAL 10

Skill Builder for Section 1 (page 59)

	257 min
−1 hr 22 min =	−82 min
	175 min
+2 hr 2 min =	+122 min
	297 min
−3 hr 13 min =	−193 min
	104 min
+2 hr 40 min =	+160 min
	264 min
−1 hr 47 min =	−107 min
	157 min
+5 hr 1 min =	+301 min
	458 min
−3 hr 19 min =	−199 min
	259 min
+4 hr 46 min =	+286 min
	545 min
−5 hr 25 min =	−325 min
	220 years

Skill Builder for Section 2 (page 61)

Ex. I
1. T
2. C
3. A
4. N
5. D
6. A
7. Z
8. E
9. Y
10. O
11. U

• GOAL 11

Skill Builder (page 65)

Across
1. qt
2. Fleetwood Mac
7. degree

Down
2. foot
3. ounce
4. customary
5. gallon
6. yard
8. mile
9. inch

• GOAL 12

Skill Builder (page 69)

1. 2 mi = 3,520 yd
2. 10,600 ft = 3,533 yd
3. 135,000 in. = 3,750 yd
4. 3,800 yd
5. 140,400 in. = 3,900 yd
6. 11,750 ft = 3,917 yd
7. 3,920 yd

• GOAL 13

Skill Builder for Section 1 (page 85)

(See table at the top of the next column.)

SI (METRIC) MEASUREMENT TABLE

BASIC UNIT OF LENGTH = meter

SHORT FORMS	RELATIONSHIPS
mm = millimeter	1 mm = 0.001 m
cm = centimeter	1 cm = 0.01 m
dm = decimeter	1 dm = 0.1 m
m = meter	1 km = 1,000 m
km = kilometer	

BASIC UNIT OF CAPACITY = liter

SHORT FORMS	RELATIONSHIPS
ml = milliliter	1 ml = 0.001 l
cl = centiliter	1 cl = 0.01 l
l = liter	
kl = kiloliter	

BASIC UNIT OF MASS = gram

SHORT FORMS	RELATIONSHIPS
mg = milligram	1 mg = 0.001 g
cg = centigram	1 cg = 0.01 g
dg = decigram	1 dg = 0.1 g
g = gram	1 kg = 1,000 g
kg = kilogram	

Skill Builder for Section 2 (page 87)

1. mm = $\frac{1}{1,000}$ m
2. cm = $\frac{1}{100}$ m
3. dm = $\frac{1}{10}$ m
4. m
5. dam = 10 m
6. hm = 100 m
7. km = 1,000 m

Skill Builder for Section 3 (page 89)

"THINK METRIC" ENTRY BLANK

Match each temperature in the left column with its correct description in the right column. Record the correct temperatures in the answer blanks in front of the descriptions.

TEMPERATURE		DESCRIPTION
23°C	23°C	a comfortable summer day
35°C	0°C	ice
100°C	43°C	the desert
0°C	100°C	boiling water
37°C	37°C	normal body temperature
43°C	4°C	a cold winter day
4°C	35°C	a hot summer day

Name ______________________

Address ______________________

• GOAL 14

Skill Builder (page 93)

1. 30 km = 30,000 m
2. 24,750 m
3. 1,950,000 mm = 1,950 m
4. 180 dam = 1,800 m
5. 9,000 dm = 900 m
6. 7 km = 700 m
7. 3,000 cm = 30 m

• GOAL 15

Skill Builder (page 97)

1–2. O
2–3. N
3–4. E

4–5. F
5–6. O
6–7. O
7–8. T

8–9. R
9–10. U
10–11. L
11–12. E
12–13. R

• GOAL 16

Skill Builder (page 101)

1. O
2. N

3. H
4. I
5. S

6. H
7. A
8. N
9. D
10. S

• GOAL 18

Skill Builder (page 119)

1. 10
2. 3
3. 8
4. 5
5. 7
6. 9
7. 6
8. 11
9. 4
10. 21

• GOAL 19

Skill Builder (page 125)

1. g
2. i
3. d
4. a
5. h
6. b
7. a
8. k
9. m

• GOAL 20

Skill Builder (page 133)

These numbers are prime (see below):

1. 47
4. 71
6. 29
9. 37
11. 7
17. 61
18. 67

	(2)	(3)	~~4~~	(5)	~~6~~	(7)	~~8~~	~~9~~	~~10~~
(11)	~~12~~	13	~~14~~	~~15~~	~~16~~	17	~~18~~	19	~~20~~
~~21~~	~~22~~	23	~~24~~	~~25~~	~~26~~	~~27~~	~~28~~	29	~~30~~
31	~~32~~	~~33~~	~~34~~	~~35~~	~~36~~	37	~~38~~	~~39~~	~~40~~
41	~~42~~	43	~~44~~	~~45~~	~~46~~	47	~~48~~	~~49~~	~~50~~
~~51~~	~~52~~	53	~~54~~	~~55~~	~~56~~	~~57~~	~~58~~	59	~~60~~
61	~~62~~	~~63~~	~~64~~	~~65~~	~~66~~	67	~~68~~	~~69~~	~~70~~
71	~~72~~	73	~~74~~	~~75~~	~~76~~	~~77~~	~~78~~	79	~~80~~
~~81~~	~~82~~	83	~~84~~	~~85~~	~~86~~	~~87~~	~~88~~	89	~~90~~
~~91~~	~~92~~	~~93~~	~~94~~	~~95~~	~~96~~	97	~~98~~	~~99~~	~~100~~

• GOAL 21

Skill Builder (page 137)

1. I
2. M
3. P
4. R
5. O
6. P
7. E
8. R

• GOAL 22

Skill Builder (page 141)

Ex. T
1. H
2. I
3. R
4. T
5. Y

6. Y
7. E
8. A
9. R
10. S

11. O
12. L
13. D

• GOAL 23

Skill Builder (page 145)

Ex. A

1. G
2. O
3. O
4. D

5. P
6. A
7. R
8. T

• GOAL 24

Skill Builder (page 149)

Ex. 4
1. 0
2. 0
3. 0

4. Y
5. E
6. A
7. R

8. O
9. L
10. D

• GOAL 25

Skill Builder (page 159)

Ex. B
1. R
2. O
3. K
4. E
5. N

6. N
7. U
8. M
9. B
10. E
11. R

ANSWER KEY TO LEARNING GUIDE PROJECTS

• PROJECT 1 (pages 21–25)

Activity 1

(See page 61)

Activity 2

The MicroOutlet
26 Claire St. New Haven, CT 06510

WORK REPORT 1 From____________________, Assistant

Sales Representative	Average Monthly Expense	Monthly Advance (Rounded Up to Next $10)
R. Langley	$416.25	**$420**
J. Willows	505.30	**510**
S. Waverly	463.77	**470**
O. Oleson	602.33	**610**
W. Seavers	495.81	**500**
T. Neperso	512.75	**520**
J. Schmidt	526.70	**530**
M. Lopez	574.70	**580**
M. Wong	486.20	**490**
F. Murphy	502.90	**510**

The MicroOutlet

EXPENSE REPORT of *Dan Jackovich* Period ending *April 30,* 19 *--*

Date	TRAVEL From	To	TRANSPORTATION Auto Mileage Daily Total	Auto Mileage Expense	Car Plan	Tolls & Parking	Car Rental	Plane & Train	Local Trans.	LIVING EXPENSE Hotel	Meals	Enter-tainment	Phone Telegraph Postage Supplies	Misc.	DAILY TOTALS	Date
1																1
2																2
3																3
4																4
5																5
6																6
7																7
8																8
9																9
10																10
11																11
12																12
13																13
14																14
15																15
16																16
17																17
18																18
19																19
20																20
21																21
22																22
23																23
24																24
25						4 25				36 20	14 15	—	2 25	—	56 85	25
26						50				32 50	11 60	—	—	—	44 60	26
27						7 00				35 00	18 35	13 40	—	4 70	78 45	27
28						1 50				37 50	14 35	—	1 97	—	55 42	28
29						2 10				—	7 00	—	3 80	—	12 90	29
30																30
31																31
Leased Car Plan																
		PERIOD TOTALS ▶				15 45				141 20	65 45	13 40	8 02	4 70	248 22 GRAND TOTAL	

The MicroOutlet

EXPENSE REPORT of *Dan Jackovich* Period ending *May 31,* 19*--*

Date	TRAVEL From	TRAVEL To	Auto Mileage Daily Total	Auto Mileage Expense	Car Plan	Tolls & Parking	Car Rental	Plane & Train	Local Trans.	Hotel	Meals	Entertainment	Phone Telegraph Postage Supplies	Misc.	DAILY TOTALS	Date	
1																1	
2																2	
3						3.25	12.50			29.50	14.75		2.05		62.05	3	
4									7.25	32.50	16.70		.85		~~57.20~~	4	57.30
5									3.75	31.75	17.35		1.35		54.20	5	
6									7.75	30.50	15.75		1.85		~~55.95~~	6	55.85
7							13.75			29.75	18.25		3.15		64.90	7	
8																8	
9																9	
10									3.55		7.55		2.55		~~12.65~~	10	13.65
11									4.15		15.75	12.50	.95		33.35	11	
12									2.60		6.50		3.25		~~12.25~~	12	12.35
13									9.75		17.75		1.85		29.35	13	
14									6.25		15.25				~~22.50~~	14	21.50
15																15	
16																16	
17							14.25				18.55	13.50	.75		~~47.55~~	17	47.05
18									3.55		11.15		2.80		17.50	18	
19									7.75		16.35		4.75		~~27.85~~	19	28.85
20						4.75					19.20		3.35		27.30	20	
21									5.65		9.95		2.95		~~18.45~~	21	18.55
22																22	
23																23	
24																24	
25																25	
26						2.50	12.95				11.70		3.05		~~31.20~~	26	30.20
27									6.95		14.35		2.65		~~23.85~~	27	23.95
28							14.65				15.10		1.15		30.90	28	
29																29	
30																30	
31																31	
Leased Car Plan																	
PERIOD TOTALS ▶						10.50	~~57.10~~ 68.10		~~69.95~~ 68.95	154.00	261.95	26.00	39.30		~~629.80~~	GRAND TOTAL	628.80

Signature *Dan Jackovich* Date *June 1, 19--*

Cash Advance *$580.00*

Monthly Expenses *~~$629.80~~ $628.80*

Difference *~~$ 49.80~~ $48.80*

☐ Owed to company ☒ Owed to employee

If expenses were greater than advance, the difference will be added to your next cash advance. If expenses were smaller, the difference will be deducted from your next cash advance.

• **PROJECT 2 (pages 47–52) Activities 1 through 6**

BANK RECONCILIATION FORM

Prepared by ____________________

____________________ **19** ______

(A) BALANCE SHOWN ON BANK STATEMENT $ 435.42

(B) BALANCE SHOWN ON CHECKBOOK STUB $ 424.98

(C) Plus: Deposits in Transit

Date	Amount	
1/30	250	00

(D) Total Deposits in Transit 250.00

(E) SUBTOTAL $ 685.42

(F) Less: Checks Outstanding

Number	Amount	
6	95	42
9	10	14
10	134	00
11	7	23
12	8	15

(G) Total Checks Outstanding 254.94

(H) ADJUSTED BANK BALANCE $ 430.48

(I) Plus: Corrections

Description	Amount	
Interest	5	50

(J) Total Additions 5.50

(K) SUBTOTAL $ 430.48

(L) Less: Charges, Fees, and Corrections

Description	Amount	
Service Charge		

(M) Total Deductions 0

(N) ADJUSTED CHECKBOOK BALANCE $ 430.48

• PROJECT 3 (pages 71–82)

Activity 1

1. program
2. PRINT
3. RUN, LIST, NEW
4. LIST
5. 10 PRINT 4556 + 3672 + 9870
   20 END
6. Line numbers
7. 10 PRINT $250 + $679 − $679
   20 END
8. ENTER or RETURN
9. RUN
10. 10 PRINT "COMPUTERS ARE FAST TIRELESS WORKERS"
    20 END

Activity 2

1. 3300
2. 10 PRINT 55 / 1000
3. 10 PRINT 20000 + 33431
   20 END
4. 10 PRINT 17 * 13
   20 END
5. 10 PRINT 999 − 33
   20 END
6. 10 PRINT 323 / 19
   20 END
7. 10 PRINT 2000 * 90001
   20 END
8. 10 PRINT 983542 / 2
   20 END

Activity 3

1. (Only lines 30 and 40 produce output.)
   COMPUTERS WORK FAST BUT THEY'RE NOT VERY BRIGHT
2. COMPUTERS WORK FAST B$
3. COMPUTERS WORK FAST
4. 10 LET A$ = "A PROGRAM TELLS THE COMPUTER EXACTLY WHAT TO DO"
   20 PRINT A$
   30 END
5. 10 LET A$ = "THIS IS WRITTEN IN"
   20 LET B$ = "A LANGUAGE CALLED BASIC"
   30 PRINT A$
   40 PRINT B$
   50 END
   (Note: answers may vary for lines 10 and 20.)

Activity 4

1. THE INTEREST IS $50
2. a. 55 REMARK *** INPUT VALUES FOR PRINCIPAL, RATE, AND TIME ***
   b. 85 REMARK *** CALCULATIONS ***
   c. 95 REMARK *** OUTPUT ***
3. THE INTEREST IS $100
4. THE INTEREST IS $20
5. THE INTEREST IS $10
6. 10 REMARK THIS PROGRAM CALCULATES INTEREST
   20 REMARK P = PRINCIPAL
   30 REMARK R = RATE
   40 REMARK T = TIME
   50 REMARK I = INTEREST
   55 REMARK *** INPUT VALUES FOR PRINCIPAL, RATE, AND TIME ***
   60 LET P = 36.90
   70 LET R = .17
   80 LET T = 2.5
   85 REMARK *** CALCULATIONS ***
   90 LET I = P * R * T
   95 REMARK *** OUTPUT ***
   100 PRINT "THE INTEREST IS"
   110 PRINT "$" I
   120 END
7. 10 REMARK THIS PROGRAM CALCULATES INTEREST
   20 REMARK P = PRINCIPAL
   30 REMARK R = RATE
   40 REMARK T = TIME
   50 REMARK I = INTEREST
   55 REMARK *** INPUT VALUES FOR PRINCIPAL, RATE, AND TIME ***
   60 LET P = 1742
   70 LET R = .11
   80 LET T = 3
   85 REMARK *** CALCULATIONS ***
   90 LET I = P * R * T
   95 REMARK *** OUTPUT ***
   100 PRINT "THE INTEREST IS"
   110 PRINT "$" I
   120 END

Activity 5

1. Lines 50, 70, and 80 print prompting messages.
   Lines 60 and 90 print question marks.
2. 45 REMARK *** INPUT VALUES FOR NUMBER SOLD AND PRICE ***
   95 REMARK ***CALCULATIONS***
   105 REMARK ***OUTPUT***
3. It lets you use the same program time and again without having to retype it, just inserting the missing facts.
4. a. HOW MANY ITEMS WERE RECEIVED TODAY?
   b. sits and waits
5. THE CURRENT BALANCE IS 18
6. THE CURRENT BALANCE IS 57
7. 10 REMARK THIS PROGRAM AVERAGES 3 TEST GRADES
   20 REMARK N$ = STUDENT NAME
   30 REMARK A = AVERAGE
   40 REMARK T1 = TEST 1
   50 REMARK T2 = TEST 2
   60 REMARK T3 = TEST 3
   65 REMARK ***INPUT VALUES FOR NAME AND TEST SCORES***
   70 PRINT "WHAT IS THE STUDENT'S NAME?"
   80 INPUT N$
   90 PRINT "WHAT IS THE SCORE ON TEST 1?"
   100 INPUT T1
   110 PRINT "WHAT IS THE SCORE ON TEST 2?"
   120 INPUT T2
   130 PRINT "WHAT IS THE SCORE ON TEST 3?"
   140 INPUT T3
   145 REMARK ***CALCULATIONS***
   150 LET A = (T1 + T2 + T3)/3
   155 REMARK ***OUTPUT***
   160 PRINT "THE AVERAGE OF 3 TEST GRADES IS"
   170 PRINT A
   180 END

• PROJECT 4 (pages 111–116) Activity 1

Form **W-4**

Department of the Treasury—Internal Revenue Service

Employee's Withholding Allowance Certificate

Print your full name ▶ | Your social security number ▶ 123 : 45 : 6789

Address (including ZIP code) ▶

Marital status: ☑ Single ☐ Married ☐ Married, but withhold at higher Single rate

Note: *If married, but legally separated, or spouse is a nonresident alien, check the single block.*

1 Total number of allowances you are claiming (from line F of the worksheet on page 2) 1

2 Additional amount, if any, you want deducted from each pay (if your employer agrees) . . . . . $ 0

3 I claim exemption from withholding because (see instructions and check boxes below that apply):

a ☐ Last year I did not owe any Federal income tax and had a right to a full refund of **ALL** income tax withheld, **AND**

b ☐ This year I do not expect to owe any Federal income tax and expect to have a right to a full refund of **ALL** income tax withheld. If both a and b apply, enter "EXEMPT" here . . . . . ▶

c If you entered "EXEMPT" on line 3b, are you a full-time student? . . . . . ☐ Yes ☐ No

Under the penalties of perjury, I certify that I am entitled to the number of withholding allowances claimed on this certificate, or if claiming exemption from withholding, that I am entitled to claim the exempt status.

Employee's signature ▶ *Student's signature* Date ▶ *Today's date* , 19

Employer's name and address (including ZIP code) **(FOR EMPLOYER'S USE ONLY)** | **Employer identification number**

Activity 2

a. Line 10
b. Line 11
c. Line 9

Activity 3

Form **1040A** Department of the Treasury—Internal Revenue Service
U.S. Individual Income Tax Return

Use IRS label. Otherwise, please print or type.

Your first name and initial (if joint return, also give spouse's name and initial) | Last name | Your social security number 123 : 45 : 6789

Present home address (Number and street, including apartment number, or rural route) | Spouse's social security no.

City, town or post office, State and ZIP code | Your occupation ▶ *Administrative assistant* | Spouse's occupation ▶

Activity 4

Presidential Election Campaign ▶ Do you want $1 to go to this fund? . . . . . Yes ☐ No
If joint return, does your spouse want $1 to go to this fund? . . . Yes ☐ No
Note: *Checking "Yes" will not increase your tax or reduce your refund.*

For Privacy Act and Paperwork Reduction Act Notice, see page 23 of Instructions

Filing Status
Check Only One Box.

1 ☑ Single
2 ☐ Married filing joint return (even if only one had income)
3 ☐ Married filing separate return. Enter spouse's social security no. above and full name here ▶
4 ☐ Head of household (with qualifying person). (See page 8 of Instructions.) If he or she is your unmarried child, enter child's name ▶

Exemptions
Always check the box labeled Yourself. Check other boxes if they apply.

5a ☑ Yourself ☐ 65 or over ☐ Blind
b ☐ Spouse ☐ 65 or over ☐ Blind
Enter number of boxes checked on 5a and b ▶ 1

c First names of your dependent children who lived with you ▶
Enter number of children listed on 5c ▶

d Other dependents:

(1) Name	(2) Relationship	(3) Number of months lived in your home.	(4) Did dependent have income of $1,000 or more?	(5) Did you provide more than one-half of dependent's support?

Enter number of other dependents ▶

6 Total number of exemptions claimed . . . . . Add numbers entered in boxes above ▶ 1

Activity 5

Please Attach Copy B of Forms W–2

7	**Wages, salaries, tips, etc.** *(Attach Forms W–2. See page 10 of Instructions)*	7	11,856 10
8a	**Interest income** . . (Complete page 2 if over $400 or you have any All-Savers interest) . . . . — 8a		
b	**Dividends** . . . . . *(Complete page 2 if over $400)* . . . . — 8b		
c	**Total** *(add lines 8a and 8b)* . . . . — 8c		
d	**Exclusion** *(See page 11 of Instructions)* . . . . — 8d		
e	**Subtract line 8d from line 8c** *(but not less than zero)*	8e	
9a	Unemployment compensation (insurance). Total received from Form(s) 1099–UC		
b	Taxable amount, if any, from worksheet on page 12 of Instructions	9b	
10	Adjusted gross income *(add lines 7, 8e, and 9b). If under $10,000, see page 13 of Instructions on "Earned Income Credit"*	10	11,856 10
11	Multiply $1,000 by the total number of exemptions claimed on line 6	11	1,000 00
12	Taxable income *(subtract line 11 from line 10)*	12	10,856 10

Activity 6

Attach Payment Here

13a	**Credit for contributions to candidates for public office.** *(See page 13 of Instructions)* . . . . — 13a		
	IF YOU WANT IRS TO FIGURE YOUR TAX, PLEASE STOP HERE AND SIGN BELOW.		
b	**Total Federal income tax withheld** *(If line 7 is more than $29,700, see page 13 of Instructions)* . . . — 13b 1,842 10		
c	**Earned income credit** *(from page 14 of Instructions)* . . . . — 13c		
14	**Total** *(add lines 13a, b, and c)*	14	1,842 10
15a	Tax on the amount on line 12. *(See page 15 of Instructions; then find your tax in the Tax Table on pages 17–22)* . . . . . — 15a		
b	Advance earned income credit (EIC) *(from Form W–2)* . . . — 15b		
16	**Total** *(add lines 15a and 15b)*	16	1,553 00
17	If line 14 is larger than line 16, enter amount to be **REFUNDED TO YOU** . . . . ▶	17	289 10
18	If line 16 is larger than line 14, enter **BALANCE DUE.** Attach check or money order for full amount payable to "Internal Revenue Service." Write your social security number and "1981 Form 1040A" on it. ▶	18	

Activity 7

Answers will vary.

Form **1040A** Department of the Treasury—Internal Revenue Service
U.S. Individual Income Tax Return

Use IRS label. Otherwise, please print or type.	Your first name and initial (if joint return, also give spouse's name and initial) ANGELINA CAPRI	Last name	Your social security number
	Present home address (Number and street, including apartment number, or rural route)		Spouse's social security no.
	City, town or post office, State and ZIP code	Your occupation ▶ / Spouse's occupation ▶	

Presidential Election Campaign ▶ Do you want $1 to go to this fund? . . . Yes / No
If joint return, does your spouse want $1 to go to this fund? . . . Yes / No
Note: *Checking "Yes" will not increase your tax or reduce your refund.*

For Privacy Act and Paperwork Reduction Act Notice, see page 23 of Instructions

Filing Status — Check Only One Box.
1 ✓ Single
2 Married filing joint return (even if only one had income)
3 Married filing separate return. Enter spouse's social security no. above and full name here ▶
4 Head of household (with qualifying person). (See page 8 of Instructions.) If he or she is your unmarried child, enter child's name ▶

Exemptions — Always check the box labeled Yourself. Check other boxes if they apply.
5a ✓ Yourself — 65 or over — Blind
b Spouse — 65 or over — Blind
Enter number of boxes checked on 5a and b ▶ 1
c First names of your dependent children who lived with you ▶
Enter number of children listed on 5c ▶

d Other dependents: (1) Name	(2) Relationship	(3) Number of months lived in your home.	(4) Did dependent have income of $1,000 or more?	(5) Did you provide more than one-half of dependent's support?

Enter number of other dependents ▶
6 Total number of exemptions claimed . . . Add numbers entered in boxes above ▶ 1

Please Attach Copy B of Forms W-2 Here

7	**Wages, salaries, tips, etc.** *(Attach Forms W-2. See page 10 of Instructions)*	7	14,873 30
8a	**Interest income** (Complete page 2 if over $400 or you have any All-Savers interest)	8a	
b	**Dividends** *(Complete page 2 if over $400)*	8b	
c	**Total** *(add lines 8a and 8b)*	8c	
d	**Exclusion** *(See page 11 of Instructions)*	8d	
e	**Subtract line 8d from line 8c** *(but not less than zero)*	8e	
9a	Unemployment compensation (insurance). Total received from Form(s) 1099-UC		
b	Taxable amount, if any, from worksheet on page 12 of Instructions	9b	
10	Adjusted gross income *(add lines 7, 8e, and 9b). If under $10,000, see page 13 of Instructions on "Earned Income Credit"*	10	14,873 30
11	Multiply $1,000 by the total number of exemptions claimed on line 6	11	1,000 00
12	Taxable income *(subtract line 11 from line 10)*	12	13,873 30
13a	**Credit for contributions to candidates for public office.** *(See page 13 of Instructions)*	13a	
	IF YOU WANT IRS TO FIGURE YOUR TAX, PLEASE STOP HERE AND SIGN BELOW.		
b	**Total Federal income tax withheld** *(If line 7 is more than $29,700, see page 13 of Instructions)*	13b	2,412 80
c	**Earned income credit** *(from page 14 of Instructions)*	13c	
14	**Total** *(add lines 13a, b, and c)*	14	2,412 80
15a	Tax on the amount on line 12. *(See page 15 of Instructions; then find your tax in the Tax Table on pages 17–22)*	15a	
b	Advance earned income credit (EIC) *(from Form W-2)*	15b	
16	**Total** *(add lines 15a and 15b)*	16	2,284 00
17	If line 14 is larger than line 16, enter amount to be **REFUNDED TO YOU** ▶	17	128 80
18	If line 16 is larger than line 14, enter **BALANCE DUE.** Attach check or money order for full amount payable to "Internal Revenue Service." Write your social security number and "1981 Form 1040A" on it. ▶	18	

Attach Payment Here

Please Sign Here
Under penalties of perjury, I declare that I have examined this return, including accompanying schedules and statements, and to the best of my knowledge and belief, it is true, correct, and complete. Declaration of preparer (other than taxpayer) is based on all information of which preparer has any knowledge.

▶ Your signature — Date ▶ Spouse's signature (if filing jointly, BOTH must sign even if only one had income)

Paid Preparer's Use Only
Preparer's signature ▶ | Date | Check if self-employed ▶ ☐ | Preparer's social security no.
Firm's name (or yours, if self-employed) and address ▶ | E.I. No. ▶ | ZIP code ▶

Form **1040A**

• PROJECT 5
(pages 127–130)

Activity 1

The MicroOutlet
26 Claire St. New Haven, CT 06510

WORK REPORT 2 From____________________, Assistant

Stock	Rate of Return per Share	Shares for $200,000		Amount of Return on Investment
		Quantity	Total Cost	
HiGro	8%	5,298	$199,999.50	$15,999.96
AmKorn	7.05%	7,920	199,980.00	14,097.60
BoMit	8.98%	6,779	199,980.50	17,964.35
TyCity	6.93%	4,878	199,998.00	13,853.52

Activity 2

Note: Assume that the lot value after the third year is the selling price.

The MicroOutlet
26 Claire St. New Haven, CT 06510

WORK REPORT 3 From____________________, Assistant

Lot	Total Price of Lot	Lot Value After 1st Year	Lot Value After 3rd Year	Profit After 3rd Year	Commission	Profit Minus Commission
Fong	$184,800	$205,128	$252,738.21	$67,938.21	$15,164.29	$52,773.92
Colter	132,675	147,269.25	181,450.45	48,775.45	10,887.03	37,888.42
Agastan	142,140	157,775.40	194,395.07	52,255.07	11,663.70	40,591.37

Activity 3

The MicroOutlet
26 Claire St. New Haven, CT 06510

WORK REPORT 4 From____________________, Assistant

Type of Bond	Annual Return (Interest)	Selling Price	Amount Purchased for $300,000	Earnings Per Year	Taxes	Return After Taxes
Midco Sanitary District	8%	$ 950	315	$25,200	$10,836.00	$14,364.00
Highline City College	14.25%	1,100	272	38,760	16,666.80	22,093.20
Western Sand and Cement	10%	790	379	37,900	16,297.00	21,603.00
Central City	15%	875	342	51,300	—	51,300.00

- **PROJECT 6**
(pages 151–156)

Activity 1

TIME CARD

Employee No. 101 Period Ending 8/31/--

Employee Name William Carnero

TIME SECTION

Day	Regular Hours	Overtime Hours
Mon.	6 1/2	
Tues.	5 1/4	
Wed.	8	3 1/4
Thurs.	6 1/2	
Fri.	8	
Sat.		4 1/2

EARNINGS SECTION

	Hours	Rate	Earnings
Regular	**34¼**	$6.80	**$232.90**
Overtime	**7¾**	$10.20	**79.05**
Total Hours	**42**	Gross Earnings	**$311.95**

TIME CARD

Employee No. 102 Period Ending 8/31/--

Employee Name Louis Capetto

TIME SECTION

Day	Regular Hours	Overtime Hours
Mon.	8	6 3/4
Tues.	8	4 1/2
Wed.	6 1/2	
Thurs.	8	5 3/4
Fri.	8	3 3/4
Sat.		

EARNINGS SECTION

	Hours	Rate	Earnings
Regular	**38½**	$6.80	**$261.80**
Overtime	**20¾**	$10.20	**211.65**
Total Hours	**59¼**	Gross Earnings	**473.45**

TIME CARD

Employee No. 103 Period Ending 8/31/--

Employee Name Elizabeth Windom

TIME SECTION

Day	Regular Hours	Overtime Hours
Mon.	8	5 1/4
Tues.	8	3 1/2
Wed.	4 1/2	
Thurs.	8	2 3/4
Fri.	8	
Sat.		3 3/4

EARNINGS SECTION

	Hours	Rate	Earnings
Regular	**36½**	$6.80	**$248.20**
Overtime	**15¼**	$10.20	**155.55**
Total Hours	**51¾**	Gross Earnings	**$403.75**

TIME CARD

Employee No. 104 Period Ending 8/31/--

Employee Name David Romanov

TIME SECTION

Day	Regular Hours	Overtime Hours
Mon.	8	2 1/2
Tues.	7 1/2	
Wed.	8	3 3/4
Thurs.	8	4
Fri.	8	3 1/2
Sat.		

EARNINGS SECTION

	Hours	Rate	Earnings
Regular	**39½**	$6.80	**$268.60**
Overtime	**13¾**	$10.20	**140.25**
Total Hours	**53¼**	Gross Earnings	**$408.85**

Activity 2

PIECE RATE PRODUCTION CARD

Employee Geraldine Smithers Period 8/31/--

Rates: Up to 100 $.20 Over 100 $.30 Part A43610

No. of Parts	Mon.	Tues.	Wed.	Thurs.	Fri.	Sat.	Total Parts	Rate	Total Pay
Up to 100	100	100	100	75	85	100	**560**	**$.20**	**$112.00**
Over 100	75	50	5	–	–	35	**165**	**$.30**	**49.50**
							Gross Earnings		**$161.50**

PIECE RATE PRODUCTION CARD

Employee Anastasia Romano Period 8/31/--

Rates: Up to 100 $.20 Over 100 $.30 Part A43610

No. of Parts	Mon.	Tues.	Wed.	Thurs.	Fri.	Sat.	Total Parts	Rate	Total Pay
Up to 100	89	100	100	94	100	100	**583**	**$.20**	**$116.60**
Over 100	–	32	27	–	21	12	**92**	**$.30**	**27.60**
							Gross Earnings		**$144.20**

PIECE RATE PRODUCTION CARD

Employee Muriel Taylor Period 8/31/--

Rates: Up to 100 $.20 Over 100 $.30 Part A43610

No. of Parts	Mon.	Tues.	Wed.	Thurs.	Fri.	Sat.	Total Parts	Rate	Total Pay
Up to 100	100	100	89	96	100	100	**585**	**$.20**	**$117.00**
Over 100	67	46	–	–	58	54	**225**	**$.30**	**67.50**
							Gross Earnings		**$184.50**

Activity 2 (continued)

PIECE RATE PRODUCTION CARD

Employee Michael Colluccio   Period 8/31/--

Rates: Up to 100 $.20   Over 100 $.30   Part A43610

No. of Parts	Mon.	Tues.	Wed.	Thurs.	Fri.	Sat.	Total Parts	Rate	Total Pay
Up to 100	100	100	100	100	100		**500**	**$.20**	**$100.00**
Over 100	62	78	44	87	56		**327**	**$.30**	**98.10**
							Gross Earnings		**$198.10**

Activity 3

1. Brennan, $636.15
   Creegan, $269.23
   Quigley, $500
   Walther, $375
2. Nieves, $1,916.67
   Lugo, $1,500
   Barnett, $3,333.33
   Saranello, $1,798.33

Activity 4

WEEKLY COMMISSION REPORT FOR PERIOD ENDING 8/31/--

Sales Representative	Straight Salary	Total Sales	Commission	Weekly Gross Earnings
Lebenthal, A.	**$100**	$5,100	**$127.50**	**$227.50**
Shears, H.	**100**	4,775	**119.38**	**219.38**
Merl, L.	**100**	5,660	**141.50**	**241.50**
Payne, T.	**100**	4,980	**124.50**	**224.50**
James, P.	**100**	5,180	**129.50**	**229.50**

Activity 5

PAYROLL REGISTER

For the Week Beginning 8/25 19-- and Ending 8/31 19-- Paid ______ 19__ Check No. ______

EMPLOYEE DATA					EARNINGS			DEDUCTIONS					
NO.	NAME	MARITAL STATUS	EXEMP.	HOURS	REGULAR	OVERTIME	TOTAL	INCOME TAX	FICA TAX	INSURANCE PREMIUMS	OTHER	TOTAL	NET PAY
30	Julia Prelid	S	1	35	230 00		230 00	34 40	15 41	3 75	6 90	**60 46**	**169 54**
31	William Carnero	M	3	35	354 50		354 50	41 70	23 75	18 75	12 00	**96 20**	**258 30**
32	Michael Flynn	S	2	40	247 00	53 40	300 40	46 60	20 13		2 00	**68 73**	**231 67**
33	Darien Morrison	S	1	35	261 40		261 40	41 60	17 51	7 75	18 30	**85 16**	**176 24**
34	Marcy Teng	M	4	35	382 50		382 50	44 40	25 63	16 90	19 12	**106 05**	**276 45**

ANSWER KEY TO LEARNING GUIDE EXTRA PRACTICES

• GOAL 1

Place Value of Whole Numbers (page 161)

1. a. ones
   b. ten thousands
   c. hundreds
   d. thousands
2. a. 6
   b. 4
   c. 0
   d. 7
3. 651,421; 24,501; 14,330; 2,450; 50
4. 53,895; 28,009; 6,315; 4,099; 2,854

Reading Whole Numbers in Words (page 161)

1. ninety-nine thousand, four hundred thirty-two
2. one thousand, four hundred one
3. four hundred seventy-five million, six hundred eighty-two thousand
4. fifty-nine million, two hundred seven
5. six thousand, seven hundred
6. fourteen million, one hundred seventy-five thousand dollars
7. thirty-nine thousand, four hundred fifty dollars
8. one thousand, two hundred ninety-five dollars
9. two hundred fifty million dollars
10. two hundred twenty-three thousand, eight hundred seventy-five

Writing Whole Numbers in Digits (page 162)

1. 700,024,051
2. 10,500,860
3. 94,670
4. 1,070,209
5. $300,000,000
6. $61,000
7. $3,700,000
8. 1,244
9. 337,000
10. $2,890,041

Place Value of Decimal Fractions (page 163)

1. a. thousandths
   b. tenths
   c. ten thousandths
   d. thousandths
2. a. 2 ten thousandths
   b. 2 hundredths
   c. 2 thousandths
   d. 2 hundredths
3. a. 4 tenths
   b. 4 thousandths

Reading Decimal Fractions in Words (page 163)

1. a. five and six hundred seventy-one thousandths
   b. ninety-eight ten thousandths
   c. seven thousand three hundred one hundred thousandths
   d. eighty-seven and four tenths
   e. eighteen hundredths
2. a. ninety-nine point eight seven
   b. six hundred fifty-three point nine nine eight three
   c. fifty-five point six
   d. eighty-eight point nine eight seven
   e. one point five

Writing Decimal Fractions in Digits (page 164)

1. .75
2. .223
3. .6
4. .99
5. .004
6. 4.64
7. .32
8. 8.023
9. 7.63
10. 10.3

Reading Common Fractions and Mixed Numbers (page 164)

1. a. two thirds
   b. twenty-five and three fourths
   c. eight hundred and one eighth
   d. fifty-five and seven eighths
   e. twenty-one and one fourth
   f. seventy-nine and one half
2. three hundred eighty-seven and one third
3. one half
4. three eighths
5. one fourth

• GOAL 2

Rounding Off Whole Numbers (page 165)

1. a. 1,000
   b. 53,000
   c. 5,000
   d. 0
2. a. 83,400
   b. 1,000
   c. 500
   d. 0
3. $185,000,000
4. $20,000

Rounding Off Decimal Fractions (page 165)

1. a. 3.250
   b. .388

c. 2.228
d. 68.046
2. a. 2.20
b. .60
c. 100.00
d. 49.21
3. 16.6%
4. 20.15%

Different Methods of Rounding (page 166)

1. a. $64.27
b. $100.10
c. $241.31
d. $18.10
2. a. $6.25
b. $.13
c. $3.46
d. $.95
3. $.50
4. 31.15 percent

• GOAL 3

Adding Columns of Numbers (page 167)

1. 25
2. 26
3. 26
4. 22
5. 25
6. 28
7. 29
8. 23
9. 16
10. 55

Carrying (page 167)

1. 156
2. 278
3. 264
4. 230
5. 254
6. 103
7. 171
8. 294
9. 101
10. 53

Adding Money Amounts (page 168)

1. $18.08
2. $15.37
3. $19.39
4. $14.96
5. $19.28
6. $28.60
7. $19.09
8. $40.35
9. $43.47
10. $26.17

Money Columns (page 168)

1. $158.97
2. $231.49
3. $179.54
4. $152.74
5. $210.51
6. $354.83
7. $90.02
8. $388.88
9. $159.71
10. $275.70

Horizontal Addition (page 169)

1. $2,776.87
2. $1,941.37
3. $838.99
4. $2,253.34
5. $1,194.62
6. $338.27
7. $2,993.19
8. $1,067.59
9. $1,940.47
10. $410.04

• GOAL 4

Borrowing (page 171)

1. 16
2. 22
3. 29
4. 39
5. 7
6. 47
7. 47
8. 37
9. 69
10. 34

Subtracting Money Amounts (page 171)

1. $284.67
2. $62.12
3. $118.61
4. $7,214.63
5. $363.27
6. $598.06
7. $0.22
8. $1,375.57
9. $1,275.45
10. $103.63

Checking Subtraction (page 171)

1. 27
2. 194
3. 28
4. 36
5. $6.83
6. 4
7. 37
8. 292
9. 58
10. $568

Long Distance Borrowing (page 172)

1. $3,064
2. $67
3. $129
4. $248,182
5. $17,665
6. $51.95
7. $273
8. $524
9. $134,500
10. $36,266

• GOAL 5

Carrying (page 173)

1. 322
2. 405
3. 216
4. 165
5. 504
6. $90
7. 112
8. 92
9. $4.30
10. $90

Multiplying Numbers Ending in Zeros (page 173)

1. 160
2. 23,000
3. 7,490
4. 7,200
5. 4,400
6. 920
7. $800
8. 3,680
9. 1,650
10. 160,000 oz

Multiplying Money Amounts (page 174)

1. $31.24
2. $598.64

3. $10.45
4. $462.00
5. $35.64
6. $18.60
7. $6.52
8. $52.00
9. $2.70
10. $76.50

Two-Digit Multipliers (page 175)

1. 22,849
2. 18,282
3. 30,375
4. 13,920
5. $107.52
6. $380.00
7. $74.25
8. $192.10
9. $780.00
10. $186.20

Multiplying by Large Numbers (page 175)

1. $31,487
2. 1,854,853
3. 1,550,143
4. $48.26
5. 50,676
6. $2,332,800
7. $3,591,204
8. $2,510,625
9. $55,755
10. 564,324

Mastering Multiplication (page 176)

1. 192,796
2. 63,114
3. 50,601
4. $20.70
5. $900.00
6. $200,000
7. $109,440
8. 325,000
9. $15,704
10. $113.40

• GOAL 6

The Meaning of Division (page 177)

1. 45 ÷ 5
2. 42 ÷ 7
3. $650 ÷ 184
4. 26 ÷ 13
5. 750 ÷ 15
6. 36 ÷ 4
7. 500 ÷ 40
8. $5,000 ÷ 900
9. 10,000 ÷ 600
10. 400 ÷ 5

Long Division (page 177)

1. 80
2. 432
3. 47
4. 82
5. 574
6. 15
7. 73
8. 220
9. 90
10. 15

Remainders and Rounding Off to the Nearest Cent (page 178)

1. $11.79
2. $8.36
3. $4.27
4. $98.41
5. $16.89
6. $55.38
7. $597.33
8. $.02
9. $358.40
10. $17.24

Finding the Average or "Mean" (page 179)

1. 39 R1
2. 21
3. 460
4. $1,187.10
5. $5.41
6. 12,420

Two-Digit Divisors (page 179)

1. 1,392 R10
2. 21 R36
3. 248
4. 504
5. 213
6. 21 R8

Long Divisors and Zeros in the Answer (page 180)

1. $1.01
2. $10.31
3. $10.72
4. $1.07
5. $125.50
6. $1,962.50

Checking Division by Multiplication (page 180)

1. okay
2. okay
3. $20.03
4. 213
5. no, $18.00
6. okay

• GOAL 7

Place Value (page 181)

1. hundredths
2. tenths
3. thousandths
4. ten thousandths
5. hundredths
6. tenths
7. thousandths
8. hundredths
9. ten thousandths
10. ten thousandths

Adding Decimal Fractions (page 181)

1. 16.02
2. 43.55
3. 146.35
4. 19.582
5. 25.9168
6. 28.1
7. 6.409
8. 219.3
9. $4.662 million
10. 16.25

Subtracting Decimal Fractions (page 182)

1. .506
2. 2.25
3. 5.277
4. 45.548
5. 1.185
6. 29.595
7. .23
8. 9.2
9. .155%
10. 29.595 thousand tons

Multiplying Decimal Fractions (page 183)

1. .0016
2. 4.125
3. 235.64
4. 17.9095
5. 6.8748
6. 38.5
7. 8.694
8. 183.26
9. $44.64 or $44.6425
10. $9.15 or $9.154

Dividing Decimal Fractions (page 184)

1. 1.23
2. .506
3. 832
4. 677
5. 2.339
6. 62
7. 23.47
8. 335
9. 62
10. $335

• GOAL 8

Writing Common Fractions (page 185)

1. a. $\frac{3}{4}$
   b. $\frac{2}{3}$
   c. $\frac{1}{5}$
   d. $\frac{8}{16}$
   e. $\frac{3}{8}$
   f. $\frac{4}{6}$
2. a. $\frac{6}{47}$
   b. $\frac{9}{28}$
   c. $\frac{22}{36}$

Common Fractions That Compare Different Numbers (page 185)

1. $\frac{55}{110}$
2. $\frac{122,000}{18,000}$
3. $\frac{444}{1,000}$
4. $\frac{700}{350}$
5. $\frac{36,000}{180,000}$
6. $\frac{6}{180}$
7. $\frac{18,000}{20,000}$
8. $\frac{150}{200}$

Decimal Fractions (page 186)

1. a. 1,179; 10,000
   b. 6; 10
   c. 43; 1,000
2. a. $\frac{454}{1,000}$
   b. $\frac{26}{100}$
   c. $\frac{1}{10,000}$
3. a. $\frac{2,000}{10,000}$; .2; or .2000
   b. $\frac{40}{100}$; .4; or .40
   c. $\frac{45}{100}$; .45
   d. $\frac{360}{1,000}$; .36; or .360

Percents (page 186)

1. a. audience
   b. employees
   c. money spent
2. a. $25
   b. $.18
   c. 27
3. c
4. c

• GOAL 9

Renaming Common Fractions as Decimal Fractions (page 187)

1. a. .3333
   b. .375
   c. .40
   d. .25
   e. .75
   f. .875
   g. .6667
   h. .80
2. a. $\frac{24}{300}$; .08
   b. $\frac{20}{125}$; .16

Renaming Decimal Fractions as Percents (page 187)

1. a. 240.7%
   b. 13%
   c. 8.4%
   d. 61.7%
   e. .42%
   f. 1,467%
   g. 305.7%
   h. 2,760%
2. a. 27%
   b. 47.1%

Renaming Percents as Decimal Fractions (page 187)

1. a. 35.8%
   b. .0414
   c. 354.05%
   d. .0019
   e. .0325
   f. 627%
   g. 34.2%
   h. 305.4%
2. a. 1.1648
   b. .11047

Renaming Common Fractions as Percents (page 188)

1. a. $\frac{2}{4}$ or $\frac{1}{2}$, .50, 50%
   b. $\frac{2}{3}$, .6667, 66.67%
   c. $\frac{1}{5}$, .20, 20%
   d. $\frac{3}{16}$, .1875, 18.75%
   e. $\frac{2}{8}$ or $\frac{1}{4}$, .25, 25%
   f. $\frac{3}{6}$ or $\frac{1}{2}$, .50, 50%
2. a. $\frac{30}{75}$, .40, 40%
   b. $\frac{5}{33}$, .1515, 15.15%
   c. $\frac{45}{200}$, .225, 22.5%

• GOAL 10

Renaming Hours and Minutes (page 189)

1. a. 134 min
   b. 271 min
   c. 118 min
   d. 87 min
2. a. 5 hr 56 min
   b. 6 hr 40 min
   c. 4 hr 43 min
   d. 5 hr 42 min
3. 95 min
4. 4 hr 43 min

Adding and Subtracting Time (page 189)

1. 7 hr 58 min
2. 17 hr 47 min

3. 12 hr 36 min
4. 10 min
5. 8 hr 32 min
6. 21 hr 44 min
7. 2 hr 25 min
8. 15 hr 45 min
9. 10 min
10. 15 hr 45 min

Multiplying and Dividing Time (page 190)

1. 15 min
2. 15 min
3. 13 min
4. 19 hr 6 min
5. $16.47
6. $92.04
7. 9 hr 20 min
8. 38 hr 45 min
9. 16 min
10. 14 hr 10 min

Calendar Facts (page 191)

1. 29
2. July and August; December and January
3. 365
4. 13
5. yes
6. 240
7. November 16
8. December 1
9. March 24
10. Friday the 21st

Calculating with Days, Months, and Years (page 192)

1. April 5
2. March 2
3. November 24
4. August 18
5. March 18
6. April 18
7. August 8
8. March 2
9. December 21
10. June 17

• GOAL 11

Short Forms for Customary Units (page 193)

1. lb
2. T
3. sq yd
4. fl oz
5. mi
6. gal
7. oz
8. pt
9. mi
10. ft

Customary Units of Length and Area (page 193)

1. b
2. d
3. c
4. c
5. a
6. a
7. b
8. a
9. b
10. a

Customary Units of Capacity (page 194)

1. c
2. a
3. c
4. b
5. c
6. a
7. c
8. a
9. c
10. b

Customary Units of Weight (page 195)

1. c
2. b
3. b
4. b
5. c
6. a
7. b
8. a
9. b
10. d

Customary Unit of Temperature (page 196)

1. c
2. b
3. c
4. a
5. c
6. c
7. c
8. a
9. d
10. a

• GOAL 12

Relationships Among Customary Units (page 197)

1. 1 gal = 4 qt
2. 16 fl oz = 1 pt
3. 3 ft = 1 yd
4. 1 pt = 16 fl oz
5. 12 in. = 1 ft
6. 1 lb = 16 oz
7. 16 oz = 1 lb
8. 1 T = 2,000 lb

Renaming Measurements (page 197)

1. 10 gal
2. 48 yd
3. 1,440 sq in.
4. 6 ft
5. 80 oz
6. 480 in.
7. 2,000 qt
8. 4 ft

Renaming More Than Once (page 198)

1. quarts, quarts
2. square yards, square yards
3. pounds, pounds
4. fluid ounces, quarts, quarts
5. 5 yd
6. 96,000 oz
7. 40
8. 64

Using a Table of Equivalent Units (page 198)

1. 4
2. 108
3. 2.5
4. 4.5
5. 400
6. 100

• GOAL 13

Common SI Units (page 199)

1. a. meter
   b. square meter
   c. gram
   d. liter
   e. degree Celsius
2. a. kilograms
   b. meters
   c. degrees Celsius
   d. liters
   e. square meters

Common SI Prefixes (page 199)

1. a. milli
   b. centi
   c. deci
2. a. smaller
   b. larger
   c. smaller
3. deciliter jars
4. 20 hectograms
5. decigram tablet
6. kilogram package

Short Forms for SI Units and Prefixes (page 199)

1. a. dal
   b. g
   c. cm
   d. kg
   e. kl
   f. °C
2. a. kilometers
   b. degrees Celsius
   c. centimeter
   d. gram
   e. deciliters
   f. millimeters

SI Units of Length (page 200)

1. c
2. b
3. a
4. d
5. a
6. b

SI Units of Capacity (page 200)

1. d
2. c
3. b
4. d
5. a
6. b

SI Units of Mass (page 200)

1. a
2. c
3. d
4. a
5. b
6. d

SI Unit of Temperature (page 201)

1. a
2. d
3. a
4. c
5. d
6. c

• GOAL 14

Renaming Measurements (page 202)

1. a. 6,000
   b. 4,000
   c. 37,000,000
   d. 17,000
   e. .2
   f. .07
2. 3,000
3. 750

Renaming More Than Once (page 202)

1. a. .047
   b. 3,000,000
   c. .2754
   d. 1,000,000
   e. 4,000,000
   f. 3,000,000
2. 10,000
3. 1,000,000

• GOAL 15

Measurement Marks on a Ruler (page 204)

1. 5 units
2. $\frac{1}{2}$ unit
3. $2\frac{1}{2}$ units
4. B
5. L
6. I
7. G
8. 1 unit
9. B
10. $3\frac{1}{2}$ units

Measuring with a Ruler (page 205)

1. A–B 1 inch, B–C $\frac{1}{2}$ inch, A–C $\frac{1}{2}$ inch
2. A–B $\frac{1}{2}$ inch, B–C $1\frac{1}{2}$ inches, A–C 2 inches
3. A–B 1 inch, B–C $\frac{1}{2}$ inch, A–C 1 inch

U.S. Customary Foot Ruler (page 205)

1. A–B $\frac{11}{16}$ inch, B–C $\frac{9}{16}$ inch, A–C $\frac{7}{8}$ inch
2. A–B 1 inch, B–C $1\frac{3}{16}$ inch, A–C $1\frac{1}{8}$ inch
3. A–B $\frac{1}{2}$ inch, B–C $\frac{13}{16}$ inch, A–C $\frac{15}{16}$ inch

SI (Metric) Centimeter Ruler (page 206)

1. A–B 27 mm, B–C 12 mm, A–C 26 mm
2. A–B 60 mm, B–C 53 mm, A–C 10 mm
3. A–B 2 mm, B–C 4 mm, A–C 4 mm

• GOAL 16

Leading Digit Estimates (page 207)

1. 1,200
2. $60.00
3. $200.00
4. 36,000
5. 400
6. 100
7. $8,000
8. $1,000.00
9. $350.00
10. $1,400

Getting Better Estimates for Addition and Subtraction (page 207)

1. b
2. b
3. b
4. a
5. c
6. c
7. d
8. b
9. c
10. b

Getting Better Estimates for Multiplication and Division (page 208)

1. a
2. b
3. b
4. c

5. c
6. d
7. b
8. b
9. b
10. a

• GOAL 17

Multiplication Tables (page 209)

1. 306
2. 228
3. 272
4. 266
5. 225
6. 13
7. 14

Sales Tax Tables (page 209)

1. a. $1.53
   b. $2.92
   c. $1.84
   d. $1.32
   e. $.74
   f. $.86
   g. $2.96
   h. $3.19
2. $2.37
3. $2.30

Withholding Tax Tables (page 209)

1. $21.60
2. $19.40
3. $17.90
4. $32.40
5. $21.40
6. $52.00
7. $44.00
8. $63.90
9. $37.40
10. $27.40

Reading a Table Inside Out (page 211)

1. a. 7 quarts
   b. 4 quarts
   c. 5 quarts
   d. 5 quarts
   e. 8 quarts
   f. 4 quarts
   g. 8 quarts
   h. 8 quarts
2. 8 quarts
3. 9 quarts

Line Graphs (page 213)

1. a. 19,000
   b. 19,800
   c. 21,000
   d. 20,000
2. December
3. August–September

Bar Graphs (page 213)

1. $60 billion
2. $200 billion
3. a. furniture, apparel
   b. package
   c. food
   d. eating and drinking

Circle Graphs (page 213)

1. subcompact, intermediate
2. compact
3. full size
4. last year
5. this year

• GOAL 18

The Meaning of Equations (page 215)

1. a. yes
   b. no
   c. yes
   d. no
2. a. Each trip = 27 ÷ 3
   b. Savings = $54 + $25 + $14
   c. Cost = $17 × 20
   d. New weight = 207 − 21

Solving Equations (page 215)

1. Rate = $\frac{\$450}{\$300}$
2. Base price = $432 − $.80
3. Rate = $\frac{\$45}{\$750}$
4. Cost = $240 + $10
5. Cash on hand = $814 + $271

Formulas (page 216)

1. Interest = Principal × Rate × Time
2. Markup = Cost price × Rate of markup
   Cost price = Markup ÷ Rate of markup
3. Increase = Base × Rate of increase
   Rate of increase = Increase ÷ Base
4. Discount = Base price × Rate of discount
5. Markup = Cost price × Rate of markup
6. Commission = Sales × Rate of commission
   Rate of commission = Commission ÷ Sales
7. Interest = Principal × Rate × Time
   Principal = Interest ÷ Rate ÷ Time
8. Percentage = Base × Rate
9. Decrease = Base × Rate of decrease
   Rate of decrease = Decrease ÷ Base
10. Commission = Sales × Rate of commission

Percentage (page 216)

1. a. 1,000
   b. 3.5%
   c. 3.975
   d. 2,000
   e. 8%
   f. 14
2. 750
3. 4.76%

Increase and Decrease (page 217)

1. a. +$1,050
   b. $3,840
   c. −5%
   d. +1.5%
   e. −$2,300
   f. $18
2. −24
3. −12.5%

Discount and Interest (page 218)

1. a. $3,400
   b. $765
   c. 15%
   d. $9,500
   e. $5,200
   f. 14%
2. $97.15
3. $3.50

Markup and Commission (page 218)

1. a. $214.00
   b. $122.58
   c. 20%
   d. 8%
   e. $950.00
   f. $202.40
2. $.44
3. $380.00

Two-Step Problems (page 219)

1. a. \$44.88; \$159.12
   b. \$3,300; \$23,300
   c. \$.19; \$1.46
   d. \$1.30; \$63.70
   e. 101.2; 193.2
   f. \$400; \$5,400
2. \$21.46
3. \$31.60

• GOAL 19

Translating the Problem (page 220)

1. a. Known: 375 cards at 80¢ each
   Unknown: earnings
   b. (3)
2. a. Known: selling price at \$2,574; investment price, \$2,000
   Unknown: profit
   b. (2)
3. a. Known: down payment of \$10,000, cost of \$80,000
   Unknown: percent of price is down payment
   b. (6)
4. a. Known: \$20,000 earnings, \$5,000 taxes
   Unknown: percent taxes are of earnings
   b. (6)
5. a. Known: \$220 earnings per week, 40 hours per week
   Unknown: hourly rate
   b. (4)

Picking the Right Information (page 220)

1. a
2. b
3. c
4. a

The T.E.S.T. Method (page 221)

1. \$8.76
2. 34 months
3. \$728
4. \$188.10
5. \$139.50

Two-Step Word Problems (page 222)

1. \$105.00
2. \$204.75
3. \$12,750
4. \$28.00
5. \$1.23

• GOAL 20

Higher and Lower Terms (page 223)

1. a. $\frac{28}{35}$
   b. $\frac{49}{56}$
   c. $\frac{21}{28}$
   d. $\frac{63}{70}$
2. a. $\frac{2}{9}$
   b. $\frac{4}{5}$
   c. $\frac{3}{18}$
   d. $\frac{3}{8}$
3. $\frac{8}{10}$
4. $\frac{5}{8}$

Simplifying or Reducing Fractions (page 224)

Answers will vary.

1. $\frac{22}{44}$ or $\frac{11}{22}$ or $\frac{1}{2}$
2. $\frac{16}{24}$ or $\frac{2}{3}$
3. $\frac{8}{12}$ or $\frac{2}{3}$
4. $\frac{81}{90}$ or $\frac{9}{10}$
5. $\frac{6}{27}$ or $\frac{2}{9}$
6. $\frac{20}{160}$ or $\frac{2}{16}$ or $\frac{1}{8}$
7. $\frac{36}{720}$ or $\frac{4}{80}$ or $\frac{1}{20}$
8. $\frac{34}{1{,}700}$ or $\frac{17}{850}$ or $\frac{1}{50}$
9. $\frac{22}{44}$ or $\frac{11}{22}$ or $\frac{1}{2}$
10. $\frac{75}{15{,}000}$ or $\frac{3}{600}$ or $\frac{1}{200}$

Simplest or Lowest Terms (page 224)

1. $\frac{2}{7}$
2. $\frac{3}{5}$
3. $\frac{5}{8}$
4. $\frac{2}{3}$
5. $\frac{3}{4}$
6. $\frac{3}{8}$
7. $\frac{7}{10}$
8. $\frac{4}{5}$
9. $\frac{4}{9}$
10. $\frac{5}{7}$

• GOAL 21

Renaming a Whole Number as a Fraction (page 225)

1. $\frac{76}{4}$
2. $\frac{130}{5}$
3. $\frac{26}{2}$
4. $\frac{460}{10}$
5. $\frac{111}{3}$
6. $\frac{120}{8}$
7. $\frac{140}{7}$
8. $\frac{360}{5}$
9. 46 halves
10. 28 quarters

Renaming a Mixed Number as an Improper Fraction (page 225)

1. $\frac{43}{10}$
2. $\frac{28}{5}$
3. $\frac{41}{4}$
4. $\frac{28}{3}$
5. $\frac{10}{2}$
6. $\frac{80}{9}$
7. $\frac{39}{8}$
8. $\frac{11}{4}$
9. 29 quarters
10. 162 fifths

Renaming a Fraction as a Mixed Number (page 225)

1. $5\frac{7}{8}$
2. $6\frac{2}{3}$
3. $5\frac{3}{5}$
4. $5\frac{7}{10}$
5. $18\frac{1}{2}$
6. $8\frac{3}{4}$
7. $2\frac{3}{6} = 2\frac{1}{2}$
8. $2\frac{5}{8}$
9. $7\frac{1}{8}$
10. $4\frac{1}{8}$

• GOAL 22

Multiplying Fractions and Whole Numbers (page 227)

1. $2\frac{7}{10}$
2. $10\frac{4}{8} = 10\frac{1}{2}$
3. $\frac{6}{40} = \frac{3}{20}$
4. $10\frac{4}{5}$
5. $\frac{15}{80} = \frac{3}{16}$
6. $2\frac{2}{3}$
7. $8\frac{2}{4} = 8\frac{1}{2}$
8. $12\frac{3}{4}$
9. $1,350
10. $68

Multiplying Mixed Numbers (page 227)

1. $4\frac{10}{15} = 4\frac{2}{3}$
2. $\frac{33}{40}$
3. $1\frac{3}{4}$
4. $10\frac{4}{10} = 10\frac{2}{5}$
5. $11\frac{1}{5}$
6. $4\frac{14}{24} = 4\frac{7}{12}$
7. $81\frac{1}{4}$
8. $131\frac{1}{4}$
9. $14\frac{1}{6}$ days
10. $22.50 or $\$22\frac{1}{2}$

Multiplying Decimal Fractions by Common Fractions (page 228)

1. $32.81
2. $145.31
3. $21.06
4. $151.88
5. $9.07
6. $158.63
7. $13.25
8. $159.47
9. $291.45
10. $14.55

• GOAL 23

Dividing with Fractions (page 229)

1. $1\frac{4}{8} = 1\frac{1}{2}$
2. 15
3. 250
4. $2\frac{7}{10}$
5. $1\frac{5}{15} = 1\frac{1}{3}$
6. $1\frac{19}{30}$
7. $46\frac{2}{3}$
8. $1\frac{1}{5}$
9. 32
10. $80,000

Dividing by Mixed Numbers (page 229)

1. $6\frac{3}{10}$
2. $\frac{210}{328} = \frac{105}{164}$
3. $58\frac{10}{11}$
4. $15\frac{39}{47}$
5. $.91 or $\$.91\frac{3}{7}$
6. $7\frac{34}{43}$
7. $1\frac{10}{30} = 1\frac{1}{3}$
8. $5\frac{5}{27}$
9. $36\frac{4}{11}$
10. $5\frac{3}{5}$

When to Multiply/When to Divide (page 230)

1. a. divide
   b. $74 \div \frac{3}{4}$
   c. 100
2. a. divide
   b. $\$1.50 \div \frac{1}{10}$
   c. $15
3. a. multiply
   b. $\$10\frac{3}{4} \times 200$
   c. $2,000
4. a. divide
   b. $\$15,000 \div \frac{1}{5}$
   c. $75,000

• GOAL 24

Adding Fractions with the Same Bottom Term (page 231)

1. $\frac{6}{8} = \frac{3}{4}$
2. $2\frac{2}{4} = 2\frac{1}{2}$
3. $2\frac{1}{3}$
4. $2\frac{3}{8}$
5. 3
6. $\frac{5}{8}$
7. $\frac{7}{10}$
8. $1\frac{2}{5}$
9. $\frac{5}{10} = \frac{1}{2}$
10. $1\frac{3}{4}$

Adding Fractions with Different Bottom Terms (page 231)

1. $1\frac{1}{8}$
2. $1\frac{3}{4}$
3. $1\frac{5}{24}$
4. $1\frac{13}{30}$
5. $\frac{7}{8}$
6. $\frac{7}{10}$
7. $\frac{5}{8}$
8. $1\frac{7}{12}$
9. $\frac{7}{8}$
10. $1\frac{27}{40}$

Adding Related Fractions (page 232)

1. $1\frac{7}{10}$
2. $1\frac{3}{8}$
3. $2\frac{1}{4}$
4. $\frac{5}{8}$
5. $1\frac{5}{10} = 1\frac{1}{2}$
6. $\frac{19}{20}$
7. $\frac{5}{6}$
8. $1\frac{11}{40}$
9. $\frac{7}{8}$
10. $1\frac{31}{40}$

Adding Mixed Numbers (page 233)

1. $3\frac{5}{8}$
2. $17\frac{1}{6}$
3. $47\frac{1}{15}$
4. $19\frac{23}{24}$
5. $12\frac{2}{10} = 12\frac{1}{5}$
6. $82\frac{7}{8}$
7. $95\frac{5}{6}$
8. $71\frac{5}{6}$
9. $68\frac{5}{12}$
10. $41\frac{4}{8} = 41\frac{1}{2}$

Adding More Than Two Fractions (page 233)

1. $8\frac{13}{15}$
2. $3\frac{17}{24}$
3. $7\frac{11}{40}$
4. $2\frac{5}{6}$
5. $3\frac{14}{15}$
6. $1\frac{33}{40}$
7. $1\frac{29}{40}$
8. $2\frac{19}{40}$
9. $2\frac{9}{20}$
10. $4\frac{1}{5}$

• GOAL 25

Subtracting Fractions (page 235)

1. $\frac{2}{4} = \frac{1}{2}$
2. $\frac{7}{15}$
3. $\frac{19}{40}$
4. $\frac{5}{24}$
5. $\frac{2}{10} = \frac{1}{5}$
6. $\frac{1}{6}$
7. $\frac{1}{40}$
8. $\frac{3}{8}$
9. $\frac{7}{30}$ carat
10. $\frac{3}{20}$ ton

Subtracting Fractions from Whole Numbers (page 235)

1. $19\frac{1}{5}$
2. $7\frac{7}{8}$
3. $9\frac{1}{4}$
4. $\frac{1}{2}$
5. $3\frac{1}{3}$
6. $9\frac{3}{10}$
7. $88\frac{1}{8}$
8. $99\frac{1}{10}$
9. $\frac{7}{8}$ oz
10. $2\frac{1}{4}$%

Subtracting Mixed Numbers (page 236)

1. $9\frac{7}{8}$
2. $3\frac{7}{12}$
3. $5\frac{27}{40}$
4. $5\frac{3}{10}$
5. $4\frac{17}{20}$
6. $4\frac{3}{4}$
7. $44\frac{2}{3}$
8. $3\frac{6}{10} = 3\frac{3}{5}$
9. $1\frac{3}{4}$ tons
10. $2\frac{2}{4}$ mpg $= 2\frac{1}{2}$ mpg

ANSWER KEY TO LEARNING GUIDE QUIZZES

• GOAL 1

(page 4)

1. d
2. b
3. b
4. c
5. b

• GOAL 2

(page 8)

1. a
2. c
3. a
4. c
5. b
6. a
7. d
8. b
9. a
10. b

• GOAL 3

Section 1
(page 12)

1. c
2. d
3. b
4. c
5. a

• GOAL 3

Section 2
(page 14)

1. d
2. a
3. a
4. a
5. b

• GOAL 4

Section 1
(page 18)

1. b
2. a
3. a
4. c
5. d
6. d
7. a
8. b
9. d
10. b

• GOAL 4

Section 2
(page 20)

1. b
2. b
3. c
4. a
5. d

• GOAL 5

Section 1
(page 28)

1. c
2. b
3. b
4. c
5. d
6. c
7. b
8. c
9. b
10. a

• GOAL 5

Section 2
(page 30)

1. a
2. b
3. c
4. b
5. d

• GOAL 6

Section 1
(page 34)

1. c
2. a
3. c
4. b
5. b

• GOAL 6

Section 2
(page 36)

1. c
2. d
3. b
4. d
5. c

• GOAL 7

Section 1
(page 40)

1. a
2. c
3. d
4. b
5. c

• GOAL 7

Section 2
(page 42)

1. a
2. b
3. c
4. b
5. c

• GOAL 8

(page 46)

1. b
2. d
3. a
4. b
5. c
6. a
7. b
8. c
9. d
10. c

• GOAL 9

(page 56)

1. d
2. c
3. d
4. a
5. c
6. b
7. c
8. d
9. b
10. b

• GOAL 10

Section 1
(page 60)

1. a
2. d
3. b
4. d
5. b

• GOAL 10

Section 2
(page 62)

1. c
2. c
3. b
4. a
5. c

• GOAL 11

(page 66)

1. f
2. c
3. b
4. a
5. e
6. c
7. c
8. a
9. b
10. d

• GOAL 12

(page 70)

1. b
2. a
3. d
4. c
5. d
6. b
7. c
8. a
9. a
10. c

• GOAL 13

Section 1
(page 86)

1. f
2. e
3. k
4. a
5. l
6. i
7. n
8. h
9. g
10. m

• GOAL 13

Section 2
(page 88)

1. d
2. c
3. b
4. b
5. c

• GOAL 13

Section 3
(page 90)

1. d
2. b
3. b
4. b
5. a

• GOAL 14

(page 94)

1. d
2. c
3. b
4. c
5. a
6. d
7. b
8. a
9. c
10. a

• GOAL 15

(page 98)

1. c
2. a
3. a
4. c
5. a

• GOAL 16

(page 102)

1. a
2. b
3. c
4. d
5. c
6. b
7. d
8. a
9. c
10. d

• GOAL 17

Section 1
(page 107)

1. b
2. c
3. d
4. c
5. a
6. d
7. d
8. a
9. d
10. c

• GOAL 17

Section 2
(page 109)

1. c
2. a
3. b
4. b
5. a
6. a
7. d
8. d
9. b
10. a

• GOAL 18

(page 120)

1. c
2. a
3. b
4. d
5. c

• GOAL 19

(page 126)

1. c
2. a
3. d
4. c
5. d

• GOAL 20

(page 134)

1. c
2. b
3. a
4. b
5. b

• GOAL 21

(page 138)

1. b
2. d
3. d
4. c
5. b

• GOAL 22

(page 142)

1. d
2. c
3. d
4. a
5. d

• GOAL 23

(page 146)

1. b
2. d
3. a
4. c
5. b

• GOAL 24

(page 150)

1. c
2. a
3. b
4. c
5. a

• GOAL 25

(page 160)

1. d
2. c
3. b
4. b
5. c

ANSWER KEY TO LEARNING GUIDE PRECHECKS, GOAL EVALUATIONS, ALTERNATE GOAL EVALUATIONS

• GOAL 1

	Precheck	Goal Evaluation	Alternate Goal Evaluation
1.	a	b	d
2.	d	b	a
3.	b	d	c
4.	b	c	d
5.	d	a	c
6.	c	d	b
7.	a	d	a
8.	d	c	b
9.	b	c	d
10.	a	b	c

• GOAL 2

	Precheck	Goal Evaluation	Alternate Goal Evaluation
1.	a	b	a
2.	d	a	c
3.	a	c	b
4.	d	b	a
5.	c	d	d
6.	b	c	c
7.	d	a	d
8.	a	d	a
9.	c	b	c
10.	a	a	a

• GOAL 3

Precheck	Goal Evaluation	Alternate Goal Evaluation
1. b	1. c	1. c
2. c	2. a	2. c
3. a	3. a	3. d
4. d	4. c	4. a
5. b	5. a	5. b
6. a	6. a	6. c
7. b	7. d	7. d
8. a	8. b	8. a
9. b	9. a	9. d
10. d	10. d	10. b

• GOAL 4

Precheck	Goal Evaluation	Alternate Goal Evaluation
1. c	1. c	1. c
2. d	2. a	2. a
3. a	3. d	3. d
4. a	4. c	4. c
5. a	5. b	5. b
6. d	6. b	6. b
7. b	7. b	7. b
8. a	8. c	8. c
9. b	9. b	9. b
10. c	10. a	10. a

• GOAL 5

Precheck	Goal Evaluation	Alternate Goal Evaluation
1. c	1. b	1. b
2. b	2. c	2. c
3. a	3. d	3. d
4. b	4. c	4. c
5. d	5. c	5. c
6. b	6. a	6. a
7. c	7. b	7. b
8. a	8. c	8. c
9. a	9. c	9. c
10. d	10. d	10. d

• GOAL 6

	Precheck	Goal Evaluation	Alternate Goal Evaluation
1.	c	b	b
2.	a	d	d
3.	b	a	a
4.	d	c	c
5.	a	c	c
6.	b	d	d
7.	d	a	a
8.	c	b	b
9.	a	c	c
10.	b	c	c

• GOAL 7

	Precheck	Goal Evaluation	Alternate Goal Evaluation
1.	c	b	b
2.	b	a	a
3.	c	d	c
4.	d	c	a
5.	b	c	b
6.	a	c	b
7.	c	b	c
8.	a	d	d
9.	d	d	a
10.	a	a	c

• GOAL 8

	Precheck	Goal Evaluation	Alternate Goal Evaluation
1.	b	b	b
2.	b	b	b
3.	a	a	a
4.	a	a	a
5.	b	b	a
6.	c	c	c
7.	c	b	c
8.	d	d	d
9.	a	a	a
10.	c	c	c

• GOAL 9

	Precheck	Goal Evaluation	Alternate Goal Evaluation
1.	b	d	b
2.	d	d	d
3.	a	c	a
4.	b	a	b
5.	c	c	c
6.	b	a	b
7.	d	d	d
8.	d	c	d
9.	c	c	c
10.	a	a	a

• GOAL 10

	Precheck	Goal Evaluation	Alternate Goal Evaluation
1.	b	b	b
2.	c	c	c
3.	c	c	c
4.	d	d	c
5.	b	b	c
6.	d	d	d
7.	b	b	b
8.	d	d	d
9.	a	d	d
10.	b	b	b

• GOAL 11

	Precheck	Goal Evaluation	Alternate Goal Evaluation
1.	c	f	e
2.	d	a	f
3.	a	g	g
4.	f	e	c
5.	e	c	d
6.	b	b	b
7.	c	c	c
8.	d	d	d
9.	d	d	d
10.	d	d	d

• GOAL 12

	Precheck	Goal Evaluation	Alternate Goal Evaluation
1.	d	d	b
2.	a	a	a
3.	c	c	d
4.	b	b	c
5.	b	b	d
6.	b	b	b
7.	c	c	c
8.	c	c	a
9.	a	b	a
10.	c	c	c

• GOAL 13

	Section 1 Precheck	Goal Evaluation	Alternate Goal Evaluation
1.	d	m	d
2.	k	d	n
3.	j	k	b
4.	a	n	j
5.	n	h	i
6.	c	j	g
7.	e	e	h
8.	b	g	k
9.	m	i	f
10.	g	c	e

	Section 2 Precheck	Goal Evaluation	Alternate Goal Evaluation
1.	b	b	b
2.	a	c	c
3.	c	d	d
4.	c	a	a
5.	d	a	a

Section 3

Precheck	Goal Evaluation	Alternate Goal Evaluation
1. a	1. b	1. b
2. c	2. b	2. b
3. d	3. c	3. a
4. c	4. b	4. b
5. b	5. d	5. a

• GOAL 14

Precheck	Goal Evaluation	Alternate Goal Evaluation
1. d	1. d	1. d
2. c	2. c	2. c
3. b	3. b	3. b
4. a	4. c	4. a
5. c	5. a	5. c
6. c	6. d	6. c
7. d	7. b	7. d
8. b	8. a	8. b
9. d	9. c	9. d
10. c	10. a	10. c

• GOAL 15

Precheck	Goal Evaluation	Alternate Goal Evaluation
1. b	1. b	1. b
2. c	2. d	2. c
3. d	3. a	3. d
4. b	4. c	4. b
5. a	5. b	5. a
6. a	6. a	6. a
7. b	7. b	7. b
8. b	8. b	8. b
9. d	9. c	9. c
10. c	10. d	10. d

• GOAL 16

Precheck	Goal Evaluation	Alternate Goal Evaluation
1. a	1. c	1. b
2. a	2. a	2. a
3. b	3. d	3. b
4. c	4. c	4. c
5. b	5. b	5. b
6. b	6. c	6. c
7. a	7. b	7. b
8. c	8. b	8. d
9. b	9. b	9. b
10. d	10. c	10. a

• GOAL 17

Precheck	Goal Evaluation	Alternate Goal Evaluation
1. d	1. d	1. d
2. d	2. d	2. d
3. a	3. a	3. d
4. a	4. a	4. a
5. b	5. b	5. b
6. c	6. c	6. c
7. b	7. b	7. b
8. c	8. c	8. b
9. d	9. d	9. d
10. b	10. b	10. b

• GOAL 18

Precheck	Goal Evaluation	Alternate Goal Evaluation
1. d	1. d	1. c
2. a	2. a	2. a
3. b	3. b	3. b
4. d	4. d	4. d
5. c	5. c	5. c
6. c	6. c	6. d
7. b	7. b	7. d
8. d	8. d	8. d
9. d	9. d	9. b
10. a	10. a	10. a

• GOAL 19

Precheck	Goal Evaluation	Alternate Goal Evaluation
1. b	1. b	1. c
2. a	2. a	2. b
3. c	3. c	3. d
4. c	4. a	4. a
5. b	5. d	5. c
6. a	6. b	6. c
7. a	7. b	7. a
8. a	8. a	8. b
9. a	9. b	9. d
10. c	10. d	10. d

• GOAL 20

Precheck	Goal Evaluation	Alternate Goal Evaluation
1. c	1. c	1. c
2. b	2. d	2. b
3. a	3. b	3. b
4. d	4. c	4. a
5. b	5. d	5. d
6. c	6. a	6. a
7. d	7. a	7. c
8. a	8. b	8. d
9. c	9. d	9. d
10. c	10. d	10. b

• GOAL 21

Precheck	Goal Evaluation	Alternate Goal Evaluation
1. a	1. c	1. a
2. c	2. d	2. c
3. b	3. a	3. b
4. b	4. d	4. b
5. a	5. a	5. a
6. c	6. d	6. c
7. b	7. a	7. b
8. d	8. c	8. b
9. b	9. b	9. d
10. b	10. b	10. a

• GOAL 22

	Precheck	Goal Evaluation	Alternate Goal Evaluation
1.	c	b	b
2.	a	d	a
3.	b	c	d
4.	d	c	d
5.	d	d	b

• GOAL 23

	Precheck	Goal Evaluation	Alternate Goal Evaluation
1.	b	c	a
2.	d	b	c
3.	a	d	c
4.	a	b	a
5.	c	c	b
6.	d	a	c
7.	c	d	d
8.	a	a	d
9.	b	c	b
10.	a	b	a

• GOAL 24

	Precheck	Goal Evaluation	Alternate Goal Evaluation
1.	a	b	c
2.	d	a	a
3.	d	d	d
4.	b	c	a
5.	c	a	c

• GOAL 25

Precheck	Goal Evaluation	Alternate Goal Evaluation
1. d	1. d	1. a
2. a	2. a	2. b
3. c	3. c	3. a
4. d	4. c	4. d
5. d	5. b	5. d
6. a	6. a	6. b
7. c	7. c	7. d
8. d	8. b	8. b
9. b	9. d	9. c
10. c	10. c	10. a

NAME ....

DATE ....

GOAL EVALUATION

GOAL 1

READING AND WRITING NUMBERS

Solve each of the following problems. Write the letter of the correct answer in the answer column at the right.

DIRECTIONS. Choose the answer that shows how each number should be read.

1. 8,001,993
  a. eight million, one hundred thousand, nine hundred ninety-three
  b. eight million, one thousand, nine hundred ninety-three
  c. eight million, nineteen hundred ninety-three
  d. eight million, ten thousand, nine hundred ninety three

2. 82.056
  a. eighty-two point fifty six thousandths
  b. eighty-two point zero five six
  c. eighty two fifty-six thousandths
  d. eighty-two and fifty-six ten thousandths

3. 409,600
  a. four hundred nine thousand six hundred
  b. forty-nine thousand six hundred
  c. forty-nine thousand, six hundred
  d. four hundred nine thousand, six hundred

4. .003
  a. three ten thousandths
  b. point three thousandths
  c. three thousandths
  d. three hundredths

DIRECTIONS. Choose the answer that shows how each number should be written in digits.

5. one million, four hundred ninety thousand
  a. 1,490,000 **b.** 1,400,090
  c. 1,000,490 **d.** 1,049,000

6. seventeen thousand and forty-two ten thousandths
  a. 170,042 **b.** 17,000.42
  c. 17,000.042 **d.** 17,000.0042

7. sixty-three and one hundredth
  a. 63.001 **b.** .6301
  c. .631 **d.** 63.01

Go on to the next page.

ANSWER COLUMN

1. ______
2. ______
3. ______
4. ______
5. ______
6. ______
7. ______

ANSWER COLUMN

8. ______
9. ______
10. ______

8. fifteen million, thirty-three

a. 15,330,000 **b.** 15,033
c. 15,000,033 **d.** 15,000,000.33

9. two thousand, three hundred seventy-five

a. 230,075 **b.** 2,300,075
c. 2,375 **d.** 2,000,375

10. fourteen ten thousandths

a. .014 **b.** .0014
c. .14 **d.** 1.04

Give the completed Goal Evaluation to your instructor.

GOAL EVALUATION

_____% ← WRITE YOUR SCORE ON YOUR TRACK RECORD

☐ NEXT GOAL
☐ Plan review

The checked box tells you what to do next.

NAME ....

DATE ....

GOAL EVALUATION

GOAL 2

ROUNDING OFF NUMBERS AND MONEY AMOUNTS

Solve each of the following problems. Write the letter of the correct answer in the answer column at the right.

DIRECTIONS. Choose the answer that shows how each number should be rounded.

1. Round off 29 to the nearest ten.
   a. 50 b. 30 c. 29 d. 20
2. Round off 4,932.7 to the nearest hundred.
   a. 4,900 b. 4,930 c. 4,000 d. 5,000
3. Round off 9,999 to the nearest thousand.
   a. 9,000 b. 9,500 c. 10,000 d. 9,900
4. Round off $1.47 to the nearest ten dollars.
   a. $1 b. $0 c. $1.40 d. $10
5. Round off 49,506 to the nearest hundred thousand.
   a. 100,000 b. 49,000 c. 50,000 d. 0
6. Round off 59.9 to the nearest hundred.
   a. 60 b. 0 c. 100 d. 59.9
7. Round off 11.567 to the nearest tenth.
   a. 11.6 b. 11.57 c. 10 d. 12
8. Round off $183.2849 to the nearest tenth of a cent.
   a. $183.30 b. $183.28 c. $183.29 d. $183.285
9. Round down $99.99 to the nearest dollar.
   a. $90 b. $99 c. $91 d. $100
10. Round up $.8001 to the nearest cent.
   a. $.81 b. $.80 c. $.90 d. $.801

ANSWER COLUMN

1. ______
2. ______
3. ______
4. ______
5. ______
6. ______
7. ______
8. ______
9. ______
10. ______

Give the completed Goal Evaluation to your instructor.

GOAL EVALUATION

_____% ← WRITE YOUR SCORE ON YOUR TRACK RECORD

☐ NEXT GOAL
☐ Plan review

The checked box tells you what to do next.

NAME ....

DATE ....

GOAL EVALUATION

GOAL 3 ADDING WHOLE NUMBERS AND MONEY AMOUNTS

Solve each of the following problems. Show all work. Write the letter of the correct answer in the answer column at the right.

•SECTION 1

DIRECTIONS. Add.

1.
```
   92
   39
   28
  +91
```
a. 2,320
b. 230
c. 250
d. 260

2.
```
   62
   49
   87
  +71
```
a. 269
b. 2,519
c. 259
d. 224

3. 32 + 48 + 71 + 57
a. 208
b. 1,918
c. 198
d. 127

4. 20 + 44 + 53 + 17
a. 1,214
b. 124
c. 134
d. 114

5. 29 + 86 + 98 + 35
a. 248
b. 2,228
c. 228
d. 230

•SECTION 2

DIRECTIONS. Add

6.
```
  $26,749
   232.99
   564.57
  +74,362
```
a. $101,908.56
b. $101,907.156
c. $101,808.56
d. $90,797.46

ANSWER COLUMN

1. ______
2. ______
3. ______
4. ______
5. ______
6. ______

Go on to the next page.

7. $21,487
   65,551
   924.48
   +691.71

   a. $86,443.19
   b. $88,653.119
   c. $88,553.19
   d. $88,654.19

ANSWER COLUMN

7. ______
8. ______
9. ______
10. ______

8. $74,112 + $265.42 + $499.83 + $67,516
   a. $131,272.25
   b. $142,393.25
   c. $142,392.125
   d. $218,153

9. $691.39 + $54,560 + $.25 + $903.38
   a. $56,155.02
   b. $56,154.102
   c. $214,052
   d. $54,054.82

10. $241.47 + $32,514 + $825.91 + $69,417
   a. $208,669
   b. $91,987.38
   c. $102,997.138
   d. $102,998.38

Give the completed Goal Evaluation to your instructor.

GOAL EVALUATION

Section 1 ______%
2 ______%
☐ NEXT GOAL
☐ Plan review

← WRITE YOUR SCORE ON YOUR TRACK RECORD

The checked box tells you what to do next.

NAME ....

DATE ....

GOAL EVALUATION

GOAL 4

SUBTRACTING WHOLE NUMBERS AND MONEY AMOUNTS

Solve each of the following problems. Show all work. Write the letter of the correct answer in the answer column at the right.

•SECTION 1

DIRECTIONS. Subtract.

1. $57 - 39$
a. 28 **b.** 22 **c.** 18 **d.** 20

2. $90 - 21$
a. 69 **b.** 71 **c.** 79 **d.** 70

3. $71 - 35$
a. 40 **b.** 46 **c.** 44 **d.** 36

4. $88 - 79$
a. 10 **b.** 11 **c.** 9 **d.** 19

5. $72 - 15$
a. 63 **b.** 57 **c.** 67 **d.** 60

•SECTION 2

DIRECTIONS. Subtract.

6. $75{,}326 - 26{,}408$
a. 48,908 **b.** 48,918 **c.** 51,122 **d.** 59,928

7. $50{,}029 - 12{,}435$
a. 27,586 **b.** 37,594 **c.** 42,414 **d.** 47,694

ANSWER COLUMN

1. ______
2. ______
3. ______
4. ______
5. ______
6. ______
7. ______

Go on to the next page.

8. $942.73 − $257.16
- **a.** $495.67
- **b.** $795.67
- **c.** $685.57
- **d.** $715.63

9. $48,727 − $307.24
- **a.** $48,418.86
- **b.** $48,419.76
- **c.** $48,420.86
- **d.** $18,003

ANSWER COLUMN

8. ______

9. ______

10. ______

10. $708.64 − $329.03
- **a.** $379.61
- **b.** $379.01
- **c.** $288.61
- **d.** $421.61

Give the completed Goal Evaluation to your instructor.

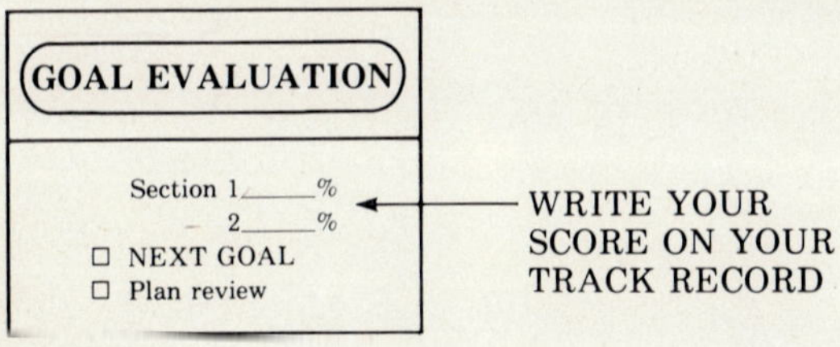

The checked box tells you what to do next.

NAME ....

DATE ....

GOAL EVALUATION

GOAL 5

MULTIPLYING WHOLE NUMBERS AND MONEY AMOUNTS

Solve each of the following problems. Show all work. Write the letter of the correct answer in the answer column at the right.

•SECTION 1

DIRECTIONS. Multiply.

1. 269 × 6

a. 1,264
b. 1,614
c. 4,864
d. 123,654

2. 832 × 5

a. 4,050
b. 5,000
c. 4,160
d. 401,510

3. 674 × 8

a. 3,662
b. 485,632
c. 11,202
d. 5,392

4. 503 × 13

a. 159
b. 2,012
c. 6,539
d. 5,539

5. 419 × 100

a. 419
b. 40,019
c. 41,900
d. 40,109

ANSWER COLUMN

1. ______
2. ______
3. ______
4. ______
5. ______

Go on to the next page.

•SECTION 2

ANSWER COLUMN

6. ______
7. ______
8. ______
9. ______
10. ______

DIRECTIONS. Multiply.

6. $44.78 × 78
- **a.** $3,492.84
- **b.** $671.70
- **c.** $3,858.84
- **d.** $3,503.84

7. $27.65 × 23
- **a.** $138.25
- **b.** $635.95
- **c.** $852.95
- **d.** $63,595

8. 9,410 × 672
- **a.** 141,950
- **b.** 6,912,070
- **c.** 6,323,520
- **d.** 5,900,070

9. 4,186 × 314
- **a.** 33,528
- **b.** 1,293,084
- **c.** 1,314,404
- **d.** 1,358,404

10. 3,658 × 354
- **a.** 43,896
- **b.** 1,148,302
- **c.** 2,084,232
- **d.** 1,294,932

Give the completed Goal Evaluation to your instructor.

GOAL EVALUATION

Section 1 ____%
2 ____% ← WRITE YOUR SCORE ON YOUR TRACK RECORD

☐ NEXT GOAL
☐ Plan review

The checked box tells you what to do next.

NAME ....

DATE ....

GOAL EVALUATION

GOAL 6

DIVIDING WHOLE NUMBERS AND MONEY AMOUNTS

Solve each of the following problems. Show all work. Write the letter of the correct answer in the answer column at the right.

•SECTION 1

DIRECTIONS. Divide and show what the remainder is, if any.

ANSWER COLUMN

1. ______
2. ______
3. ______
4. ______
5. ______

1. $7\overline{)5,909}$

a. 85 R4
b. 844 R1
c. 854
d. 850 R4

2. $5\overline{)\$48.24}$

a. \$11.01
b. \$96.44
c. \$9.60 R24
d. \$9.64 R4

3. $8\overline{)8,873}$

a. 1,109 R1
b. 119 R1
c. 11,091
d. 1,191

4. $6\overline{)\$44.08}$

a. \$73.44
b. \$7.35 R8
c. \$7.34 R4
d. \$7.35

5. $7\overline{)7,032}$

a. 104 R4
b. 1,400 R4
c. 1,004 R4
d. 14 R4

Go on to the next page.

•SECTION 2

DIRECTIONS. Divide and show what the remainder is, if any.

ANSWER COLUMN

6. ______

7. ______

8. ______

9. ______

10. ______

6. $258\overline{)78{,}848}$

a. 350 R158
b. 305 R258
c. 35 R158
d. 305 R158

7. $121\overline{)\$298.26}$

a. $2.46 R60
b. $246.60
c. $247 R39
d. $246 R60

8. $36\overline{)\$183.38}$

a. $509.14
b. $5.09 R14
c. $5.90 R14
d. $5.9 R14

9. $87\overline{)52{,}906}$

a. 68 R10
b. 60 R76
c. 608 R10
d. 680 R10

10. $638\overline{)62{,}201}$

a. 97 R215
b. 99 R39
c. 97 R315
d. 97 R415

Give the completed Goal Evaluation to your instructor.

GOAL EVALUATION

Section 1 ______%
2 ______% ← WRITE YOUR SCORE ON YOUR TRACK RECORD

☐ NEXT GOAL
☐ Plan review

The checked box tells you what to do next.

NAME ....

DATE ....

GOAL EVALUATION

GOAL 7

PLACING THE DECIMAL POINT

Solve each of the following problems. Show all work. Write the letter of the correct answer in the answer column at the right.

•SECTION 1

DIRECTIONS. Rewrite each problem and solve.

1. 1.981 + 2.7 + 3 + 6.54
- **a.** 2.665
- **b.** 14.221
- **c.** 12.2221
- **d.** 1.4221

2. 54.7 − .056
- **a.** 54.644
- **b.** 49.1
- **c.** 54.756
- **d.** 54.744

3. .67 + 5.32 + 3.1 + 1.276
- **a.** 1.0366
- **b.** 1.906
- **c.** 9.1366
- **d.** 10.366

4. 492 − 37.6
- **a.** 116
- **b.** 455.6
- **c.** 454.4
- **d.** 455.4

5. 5 − 3.999
- **a.** 2.999
- **b.** 2.001
- **c.** 1.001
- **d.** 1.111

ANSWER COLUMN

1. ______

2. ______

3. ______

4. ______

5. ______

Go on to the next page.

•SECTION 2

ANSWER COLUMN

6. ______
7. ______
8. ______
9. ______
10. ______

Rewrite each problem and solve.

6. 2.22 × .5
a. 11.1
b. 1,110
c. 1.11
d. 10.11

7. 3.62 × 4.2
a. 12.124
b. 15.204
c. 1.5204
d. 15,204

8. 113 ÷ .4
a. 2.825
b. 28.25
c. 281
d. 282.5

9. 132 ÷ .004
a. 33
b. 330
c. 3,300
d. 33,000

10. 48.78 ÷ 1.5
a. 32.52
b. 3.252
c. 32.53
d. 48.176

Give the completed Goal Evaluation to your instructor.

GOAL EVALUATION

Section 1______%
2______% ← WRITE YOUR SCORE ON YOUR TRACK RECORD
☐ NEXT GOAL
☐ Plan review

The checked box tells you what to do next.

NAME ....

DATE ....

GOAL EVALUATION

GOAL 8 **WRITING COMMON FRACTIONS, DECIMAL FRACTIONS, AND PERCENTS**

Solve each of the following problems. Show all work. Write the letter of the correct answer in the answer column at the right.

DIRECTIONS. Choose the number that makes each statement true.

1. __?__ of the diagram is shaded.

a. $\frac{2}{6}$ **b.** $\frac{6}{8}$

c. $\frac{8}{6}$ **d.** $\frac{6}{2}$

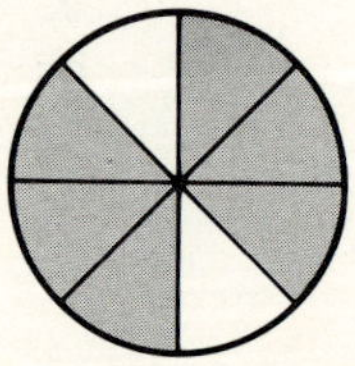

2. Cohen Office Supplies had \$12 million sales last year and \$14 million sales this year. This year's sales was __?__ of last year's sales.

a. $\frac{12}{14}$ **b.** $\frac{14}{12}$

c. $\frac{2}{14}$ **d.** $\frac{2}{12}$

3. The new computerized office in JVL Co. allows workers to process 12 cases for every 9 processed before. The number of cases being processed now is __?__ of what it was before.

a. $\frac{12}{9}$ **b.** $\frac{21}{12}$

c. $\frac{9}{12}$ **d.** $\frac{21}{9}$

4. The decimal fraction .71 means 71 out of __?__.

a. 100 **b.** 1,000

c. .10 **d.** .1

5. The decimal fraction .7 means the same thing as the common fraction __?__.

a. $\frac{7}{100}$ **b.** $\frac{7}{10}$

c. $\frac{7}{1}$ **d.** $\frac{1}{7}$

ANSWER COLUMN

1. ______

2. ______

3. ______

4. ______

5. ______

Go on to the next page.

ANSWER COLUMN

6. ______
7. ______
8. ______
9. ______
10. ______

DIRECTIONS. Choose the number that makes each statement true.

6. __?__ of the diagram is shaded.

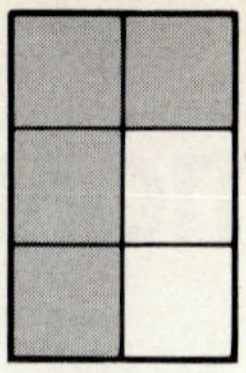

a. $\frac{4}{10}$ **b.** $\frac{6}{10}$
c. $\frac{4}{6}$ **d.** $\frac{6}{4}$

7. The decimal fraction .86 means the same thing as the common fraction __?__.

a. $\frac{.86}{100}$ **b.** $\frac{86}{100}$
c. $\frac{86}{1,000}$ **d.** $\frac{86}{10,000}$

8. 21% means __?__ parts in every 100.

a. 2.1 **b.** .21
c. 210 **d.** 21

9. If 100% of your $613 investment was lost, __?__ is left?

a. $0 **b.** $613
c. $513 **d.** $61.30

10. Assets were 150% of liabilities. The base of comparison in this statement is __?__.

a. assets **b.** assets + liabilities
c. liabilities **d.** assets − liabilities

Give the completed Goal Evaluation to your instructor.

GOAL EVALUATION

____% ← WRITE YOUR SCORE ON YOUR TRACK RECORD

☐ NEXT GOAL
☐ Plan review

The checked box tells you what to do next.

NAME ....

DATE ....

GOAL EVALUATION

GOAL 9 **RENAMING COMMON FRACTIONS, DECIMAL FRACTIONS, AND PERCENTS**

Solve each of the following problems. Show all work. Write the letter of the correct answer in the answer column at the right.

DIRECTIONS. Choose the number that makes the statement true.

1. $\frac{1}{5}$ = __?__
a. .15
b. 1.5
c. .5
d. .20

2. $\frac{3}{8}$ = __?__
a. .38
b. .267
c. 3.75
d. .375

3. $67\frac{1}{2}$% = __?__
a. 6.75
b. 67.5
c. .675
d. .0675

4. 43% = __?__
a. .43
b. 43
c. 4.3
d. .043

5. .371 = __?__
a. .371%
b. 371%
c. 37.1%
d. 3.71%

6. 2.54% = __?__
a. .0254
b. 2.54
c. .254
d. .00254

7. $5\frac{7}{8}$ = __?__
a. 5.78
b. 5.143
c. 5.87
d. 5.875

8. $2\frac{1}{4}$ = __?__
a. 2.14
b. 2.4
c. 2.25
d. 2.41

9. 7.667 = __?__
a. 76.67%
b. 7.667%
c. 766.7%
d. .7667%

10. 2.15% = __?__
a. .0215
b. 2.15
c. .00215
d. .215

ANSWER COLUMN

1. ______
2. ______
3. ______
4. ______
5. ______
6. ______
7. ______
8. ______
9. ______
10. ______

Give the completed Goal Evaluation to your instructor.

GOAL EVALUATION

_____% ← WRITE YOUR SCORE ON YOUR TRACK RECORD

☐ NEXT GOAL
☐ Plan review

The checked box tells you what to do next.

NAME ....

DATE ....

GOAL EVALUATION

GOAL 10 **WORKING WITH TIME**

Solve each of the following problems. Show all work. Write the letter of the correct answer in the answer column at the right.

•SECTION 1

DIRECTIONS. Choose the number that makes each statement true.

ANSWER COLUMN

1. ______
2. ______
3. ______
4. ______
5. ______

1. 7 hr 22 min = __?__
- **a.** 722 min
- **b.** 442 min
- **c.** 372 min
- **d.** 862 min

2. 619 min = __?__
- **a.** 25 hr 19 min
- **b.** 12 hr 19 min
- **c.** 10 hr 19 min
- **d.** 6 hr 19 min

3. 23 hr 8 min − 12 hr 52 min = __?__
- **a.** 11 hr 44 min
- **b.** 11 hr 16 min
- **c.** 10 hr 16 min
- **d.** 9 hr 16 min

4. 34 hr 23 min − 21 hr 57 min = __?__
- **a.** 14 hr 26 min
- **b.** 13 hr 34 min
- **c.** 13 hr 26 min
- **d.** 12 hr 26 min

5. 9 hr 18 min + 12 hr + 21 hr 43 min + 25 min = __?__
- **a.** 42 hr 86 min
- **b.** 43 hr 26 min
- **c.** 43 hr 86 min
- **d.** 22 hr 57 min

Go on to the next page.

•SECTION 2

ANSWER COLUMN

DIRECTIONS. Find the correct time or date.

6. How many days are there from April 9 to June 2?
a. 56 days
b. 53 days
c. 55 days
d. 54 days

7. What is the date 26 days after April 7?
a. May 2
b. May 3
c. May 1
d. April 30

8. What is the date 62 days after February 22, 1992?
a. April 25
b. April 23
c. April 26
d. April 24

9. How many days are there from April 7 to May 20?
a. 42 days
b. 45 days
c. 44 days
d. 43 days

10. What is the date 8 weeks after August 25?
a. October 21
b. October 20
c. October 19
d. October 18

6. ______
7. ______
8. ______
9. ______
10. ______

Give the completed Goal Evaluation to your instructor.

GOAL EVALUATION

Section 1 ____%
2 ____% ← WRITE YOUR SCORE ON YOUR TRACK RECORD
☐ NEXT GOAL
☐ Plan review

The checked box tells you what to do next.

NAME ....

DATE ....

GOAL EVALUATION

GOAL 11

IDENTIFYING U.S. CUSTOMARY UNITS

Solve each of the following problems. Show all work. Write the letter of the correct answer in the answer column at the right.

DIRECTIONS. Match the terms in Column I with what they mean or measure in Column II.

Column I	Column II
1. 5,280 feet	**a.** weight
2. ounce	**b.** temperature
3. inch	**c.** capacity or volume
4. 32 fluid ounces	**d.** energy
5. gallon	**e.** quart
	f. mile
	g. length

DIRECTIONS. Choose the unit that is closest in size to the object named.

6. The area of a desk
- **a.** 5 square yards
- **b.** 5 square feet
- **c.** 5 feet
- **d.** 5 square inches

7. The length of an adult's belt
- **a.** 1 foot
- **b.** 1 square yard
- **c.** 1 yard
- **d.** 1 square foot

8. The quantity of water needed to fill a sink
- **a.** 2 fluid ounces
- **b.** 2 pints
- **c.** 2 quarts
- **d.** 2 gallons

9. The weight of a quart container of juice
- **a.** 2 pints
- **b.** 2 ounces
- **c.** 2 tons
- **d.** 2 pounds

10. The width of an LP record
- **a.** 12 feet
- **b.** 12 miles
- **c.** 12 yards
- **d.** 12 inches

ANSWER COLUMN

1. ______
2. ______
3. ______
4. ______
5. ______
6. ______
7. ______
8. ______
9. ______
10. ______

Give the completed Goal Evaluation to your instructor.

GOAL EVALUATION

_____% ← WRITE YOUR SCORE ON YOUR TRACK RECORD

☐ NEXT GOAL
☐ Plan review

The checked box tells you what to do next.

NAME ....

DATE ....

GOAL EVALUATION

GOAL 12 RENAMING U.S. CUSTOMARY UNITS

Solve each of the following problems. Show all work. Write the letter of the correct answer in the answer column at the right.

DIRECTIONS. Choose the number that makes each statement true.

1. ___?___ inches = 15 feet
- **a.** 5
- **b.** 60
- **c.** 45
- **d.** 180

2. 192 fluid ounces = ___?___ pints
- **a.** 12
- **b.** 6
- **c.** 16
- **d.** 3

3. ___?___ tons = 28,000 pounds
- **a.** $2{,}333\frac{1}{3}$
- **b.** 1,750
- **c.** 14
- **d.** $777\frac{7}{9}$

4. 72 yards = ___?___ feet
- **a.** 6
- **b.** 216
- **c.** 2
- **d.** 864

5. 1,200 square inches = ___?___ square feet
- **a.** 100
- **b.** $8\frac{1}{3}$
- **c.** $33\frac{1}{3}$
- **d.** $66\frac{2}{3}$

6. 135 pounds = ___?___ ounces
- **a.** 1,620
- **b.** 2,160
- **c.** 45
- **d.** 4,860

7. 90,000 pounds = ___?___ tons
- **a.** 7,500
- **b.** 5,625
- **c.** 45
- **d.** 90

ANSWER COLUMN

1. ______
2. ______
3. ______
4. ______
5. ______
6. ______
7. ______

Go on to the next page.

8. 26,400 yards = ___?___ miles
   a. 2,200
   b. $733\frac{1}{3}$
   c. 15
   d. 5

9. 4 miles = ___?___ feet
   a. 7,040
   b. 21,120
   c. 440
   d. 1,320

ANSWER COLUMN

8. ______

9. ______

10. ______

10. 48,000 fluid ounces = ___?___ pints
    a. 375
    b. 1,500
    c. 3,000
    d. 768,000

Give the completed Goal Evaluation to your instructor.

GOAL EVALUATION

____% ← WRITE YOUR SCORE ON YOUR TRACK RECORD

☐ NEXT GOAL
☐ Plan review

The checked box tells you what to do next.

NAME ....

DATE ....

GOAL EVALUATION

GOAL 13 **IDENTIFYING SI UNITS**

Solve each of the following problems. Write the letter of the correct answer in the answer column at the right.

•SECTION 1

DIRECTIONS. Match the terms in Column I with what they mean or measure in Column II.

Column I	Column II
1. a person's height	**a.** degrees Celsius
2. a barrel's capacity	**b.** 10
3. a burger's weight	**c.** $\frac{1}{1,000}$ or .001
4. centi	**d.** liters
5. kilo	**e.** milligram
6. hectoliters	**f.** ergs
7. mg	**g.** $\frac{1}{10}$ or .1
8. deci	**h.** 1,000
9. centimeter	**i.** cm
10. milli	**j.** hl
	k. grams
	l. 100
	m. meters
	n. $\frac{1}{100}$ or .01

ANSWER COLUMN

1. ______
2. ______
3. ______
4. ______
5. ______
6. ______
7. ______
8. ______
9. ______
10. ______

•SECTION 2

DIRECTIONS. Choose the unit that is closest in size to the object named.

1. The length of a person's finger is about
a. 10 millimeters **b.** 10 centimeters
c. 10 meters **d.** 10 dekameters

2. A glass of milk contains about
a. 300 liters **b.** 3 milliliters
c. 300 milliliters **d.** 3 liters

3. The distance a car travels in 1 hour on a highway can be
a. 80 centimeters **b.** 8,000 millimeters
c. 8,000 meters **d.** 80 kilometers

4. A punch bowl can contain
a. 5 liters **b.** 5 hectoliters
c. 50 milliliters **d.** 5 centiliters

1. ______
2. ______
3. ______
4. ______

Go on to the next page.

5. The height of a school's flag pole can be
   a. 30 meters   b. 300 millimeters
   c. 3 kilometers   d. 3 decimeters

ANSWER COLUMN

5. ______

•SECTION 3

DIRECTIONS. Choose the unit that is closest in size to the object named.

1. A package of frozen peas weighs about
   a. 50 kilograms   b. 500 grams
   c. 50 milligrams   d. 500 centigrams

2. The temperature on a hot summer day can be
   a. 150°C   b. 38°C
   c. 93°C   d. 10°C

3. A human baby weighs about
   a. 45 kilograms   b. 4.5 milligrams
   c. 4.5 kilograms   d. 45 grams

4. The temperature in a kitchen freezer is about
   a. 20°C   b. −10°C
   c. 10°C   d. −50°C

5. The quantity of medicine in pill form can be
   a. 30 grams   b. 30 kilograms
   c. 30 hectograms   d. 30 milligrams

1. ______
2. ______
3. ______
4. ______
5. ______

Give the completed Goal Evaluation to your instructor.

GOAL EVALUATION

Section 1 ____%
2 ____% ← WRITE YOUR SCORE ON YOUR TRACK RECORD
3 ____%

☐ NEXT GOAL
☐ Plan review

The checked box tells you what to do next.

NAME ____________

DATE ____________

GOAL EVALUATION

GOAL 14 **RENAMING SI UNITS**

Solve each of the following problems. Show all work. Write the letter of the correct answer in the answer column at the right.

DIRECTIONS. Choose the number that makes each statement true.

1. 4.1 liters = ___?___ milliliters
- **a.** .0041
- **b.** 41
- **c.** 410
- **d.** 4,100

2. 74 dekameters = ___?___ meters
- **a.** 7.4
- **b.** .74
- **c.** 740
- **d.** 7,400

3. 9 kilograms = ___?___ grams
- **a.** .009
- **b.** 9,000
- **c.** 900
- **d.** .9

4. 98 liters = ___?___ deciliters
- **a.** 9.8
- **b.** .98
- **c.** 980
- **d.** 9,800

5. 99 hectograms = ___?___ grams
- **a.** 9,900
- **b.** 99,000
- **c.** .99
- **d.** 9.9

6. 185 centimeters = ___?___ meters
- **a.** 18.5
- **b.** 18,500
- **c.** 1,850
- **d.** 1.85

7. ___?___ grams = 852 decigrams
- **a.** 8.52
- **b.** 85.2
- **c.** 8,520
- **d.** 85,200

8. ___?___ meters = 254 dekameters
- **a.** 2,540
- **b.** 25,400
- **c.** 25.4
- **d.** .254

ANSWER COLUMN

1. ______
2. ______
3. ______
4. ______
5. ______
6. ______
7. ______
8. ______

Go on to the next page.

9. ___?___ liters = 36.1 deciliters

a. 3,610
b. 361
c. 3.61
d. .361

10. ___?___ meters = 8.17 kilometers

a. 8,170
b. 817
c. 81.7
d. .00817

ANSWER COLUMN

9. ______
10. ______

Give the completed Goal Evaluation to your instructor.

GOAL EVALUATION

____% ← WRITE YOUR SCORE ON YOUR TRACK RECORD

☐ NEXT GOAL
☐ Plan review

The checked box tells you what to do next.

NAME ....

DATE ....

GOAL EVALUATION

GOAL 15

MEASURING WITH U.S. CUSTOMARY AND SI RULERS

Solve each of the following problems. Write the letter of the correct answer in the answer column at the right.

DIRECTIONS. Use a standard inch ruler to measure the distance between the marks on each figure to the nearest $\frac{1}{16}$ inch.

Note: Measure from the left side of one dot to the left side of the next dot.

ANSWER COLUMN

1. ______
2. ______
3. ______
4. ______
5. ______

1.

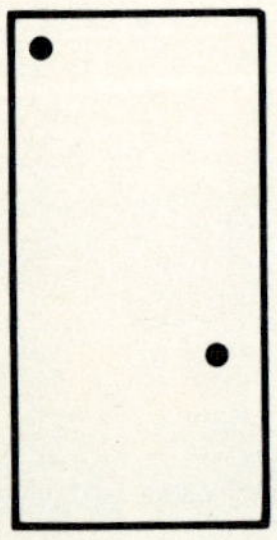

a. $\frac{13}{16}$ in.
b. $\frac{15}{16}$ in.
c. 1 in.
d. $\frac{7}{8}$ in.

2.

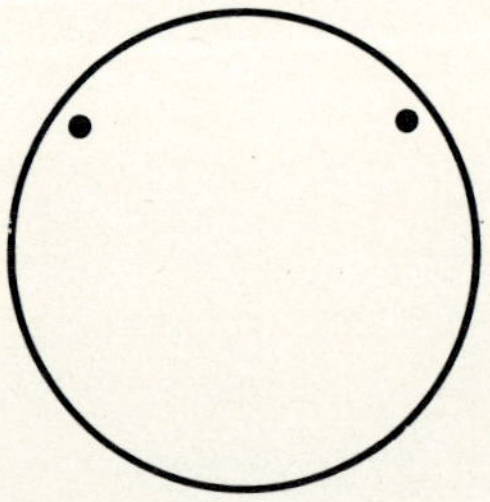

a. 1 in.
b. $\frac{13}{16}$ in.
c. $\frac{15}{16}$ in.
d. $\frac{7}{8}$ in.

3.

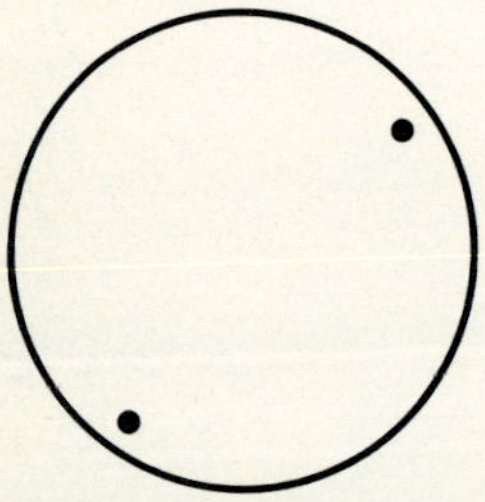

a. $1\frac{1}{16}$ in.
b. $1\frac{1}{8}$ in.
c. 1 in.
d. $1\frac{3}{16}$ in.

4.

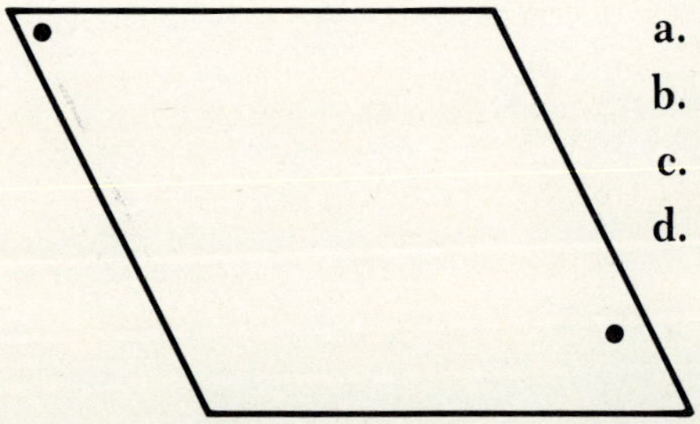

a. $2\frac{1}{16}$ in.
b. $2\frac{1}{8}$ in.
c. $2\frac{3}{16}$ in.
d. $2\frac{1}{4}$ in.

5.

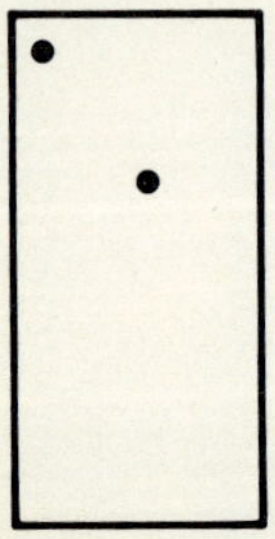

a. $\frac{9}{16}$ in.
b. $\frac{7}{16}$ in.
c. $\frac{1}{2}$ in.
d. $\frac{3}{8}$ in.

Go on to the next page.

DIRECTIONS. Use a standard centimeter ruler to measure the distance between the marks on each figure to the nearest millimeter.

Note: Measure from the left side of one dot to the left side of the next dot.

ANSWER COLUMN

6. ______
7. ______
8. ______
9. ______
10. ______

6.

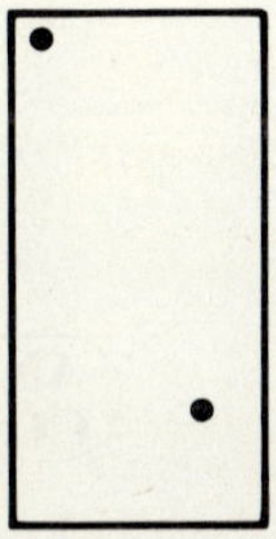

a. 27 mm
b. 28 mm
c. 25 mm
d. 29 mm

7.

a. 18 mm
b. 17 mm
c. 16 mm
d. 15 mm

8.

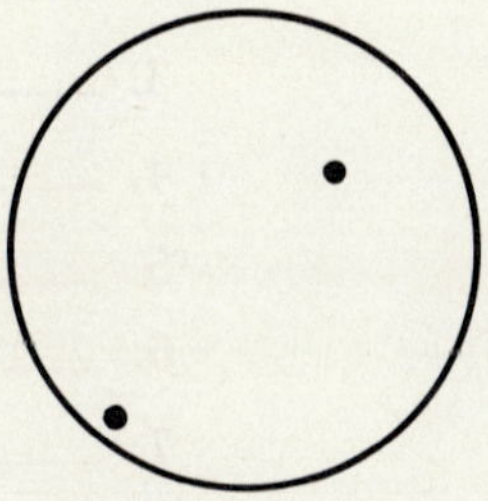

a. 24 mm
b. 22 mm
c. 25 mm
d. 21 mm

9.

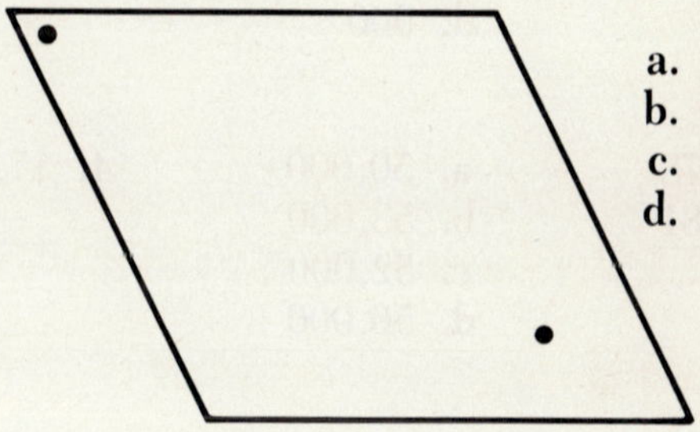

a. 48 mm
b. 47 mm
c. 49 mm
d. 46 mm

10.

a. 30 mm
b. 32 mm
c. 33 mm
d. 31 mm

Give the completed Goal Evaluation to your instructor.

GOAL EVALUATION

_____% ← WRITE YOUR SCORE ON YOUR TRACK RECORD

☐ NEXT GOAL
☐ Plan review

The checked box tells you what to do next.

NAME ....

DATE ....

GOAL EVALUATION

GOAL 16 **ESTIMATING ANSWERS**

Solve each of the following problems. Show all work. Write the letter of the correct answer in the answer column at the right.

DIRECTIONS. Estimate the answer to each problem. Don't find the exact answer. Choose the estimate that seems most reasonable.

1. 594 + 407 + 48 + 89

- **a.** 1,500
- **b.** 1,300
- **c.** 1,100
- **d.** 900

2. \$468.17 + 331.94 + 87.56 + 11.07

- **a.** \$900
- **b.** \$700
- **c.** \$1,300
- **d.** \$1,000

3. 64,378 − 7,881

- **a.** 50,000
- **b.** 53,000
- **c.** 52,000
- **d.** 56,000

4. \$7,907 − 612

- **a.** \$6,500
- **b.** \$7,000
- **c.** \$7,300
- **d.** \$7,400

5. Wanda Perez serviced 244 computers last month. There were 89 still on warranty. The owners paid for the service for the rest. How many service calls were paid for by the owners?

a. 100 **b.** 150 **c.** 160 **d.** 120

6. 25,899 × 412

- **a.** 100,000
- **b.** 1,000,000
- **c.** 10,000,000
- **d.** 100,000,000

7. 571 × 551

- **a.** 30,000
- **b.** 300,000
- **c.** 3,000,000
- **d.** 30,000,000

8. 217)2,298,215

- **a.** 1,000
- **b.** 10,000
- **c.** 100,000
- **d.** 1,000,000

9. 92)827,469

- **a.** 900
- **b.** 9,000
- **c.** 90,000
- **d.** 900,000

10. Lovelace Electronics sold 579 computers for \$376,812. What was the average price of each computer?

a. \$60,000 **b.** \$6,000 **c.** \$600 **d.** \$60

ANSWER COLUMN

1. ______
2. ______
3. ______
4. ______
5. ______
6. ______
7. ______
8. ______
9. ______
10. ______

Give the completed Goal Evaluation to your instructor.

GOAL EVALUATION

____% ← WRITE YOUR SCORE ON YOUR TRACK RECORD

☐ NEXT GOAL
☐ Plan review

The checked box tells you what to do next.

NAME ....

DATE ....

GOAL EVALUATION

GOAL 17 **READING TABLES AND GRAPHS**

Solve each of the following problems. Show all work. Write the letter of the correct answer in the answer column at the right.

•SECTION 1

DIRECTIONS. Use the tables to answer these questions.

1. According to Table 1, what is the tax on $.55?
a. $.60 **b.** $.12
c. $.04 **d.** $.05

2. According to Table 1, what is the tax on $1.81?
a. $.13 **b.** $.07
c. $.15 **d.** $.14

3. Refer to Table 2 (page 126). What is the tax on weekly pay of $180 if 3 allowances are claimed?
a. $15.90 **b.** $12.20
c. $17.80 **d.** $27.00

4. Refer to Table 2 (page 126). What is the tax on weekly pay of $155 if 3 allowances are claimed?
a. $10.20 **b.** $23.00
c. $21.10 **d.** $12.10

TABLE 1 8% SALES TAX TABLE

Amount of Sale	Tax
$ .01 to $ .10	$.00
.11 to .17	.01
.18 to .29	.02
.30 to .42	.03
.43 to .54	.04
.55 to .67	.05
.68 to .79	.06
.80 to .92	.07
.93 to 1.06	.08
1.07 to 1.18	.09
1.19 to 1.31	.10
1.32 to 1.43	.11
1.44 to 1.56	.12
1.57 to 1.68	.13
1.69 to 1.81	.14
1.82 to 1.93	.15
1.94 to 2.06	.16
2.07 to 2.18	.17
2.19 to 2.31	.18
2.32 to 2.43	.19

ANSWER COLUMN

1. ______
2. ______
3. ______
4. ______
5. ______
6. ______

•SECTION 2

DIRECTIONS. Use the graphs on pages 126 and 127 to answer these questions.

5. Refer to the graph of Merlin's Home Computers. How many microcomputers were sold in October?
a. 23,000 **b.** 23,200
c. 23,800 **d.** 24,000

6. Refer to the graph of Merlin's Home Computers. In which month were sales 21,000?
a. June **b.** May
c. July **d.** August

Go on to the next page.

TABLE 2

SINGLE Persons — **WEEKLY** Payroll Period

And the wages are—		And the number of withholding allowances claimed is—										
At least	But less than	0	1	2	3	4	5	6	7	8	9	10
		The amount of income tax to be withheld shall be—										
$135	$140	$17.80	$14.10	$10.50	$7.40	$4.70	$2.00	$0	$0	$0	$0	$0
140	145	18.70	15.10	11.40	8.20	5.40	2.70	0	0	0	0	0
145	150	19.70	16.00	12.40	9.00	6.10	3.40	.70	0	0	0	0
150	160	21.10	17.50	13.80	10.20	7.20	4.40	1.70	0	0	0	0
160	170	23.00	19.40	15.70	12.10	8.80	5.80	3.10	.40	0	0	0
170	180	24.90	21.30	17.60	14.00	10.40	7.30	4.50	1.80	0	0	0
180	190	27.00	23.20	19.50	15.90	12.20	8.90	5.90	3.20	.50	0	0
190	200	29.40	25.10	21.40	17.80	14.10	10.50	7.40	4.60	1.90	0	0
200	210	31.80	27.20	23.30	19.70	16.00	12.40	9.00	6.00	3.30	.60	0
210	220	34.20	29.60	25.20	21.60	17.90	14.30	10.60	7.50	4.70	2.00	0
220	230	36.60	32.00	27.40	23.50	19.80	16.20	12.50	9.10	6.10	3.40	.80
230	240	39.00	34.40	29.80	25.40	21.70	18.10	14.40	10.70	7.70	4.80	2.20
240	250	41.40	36.80	32.20	27.60	23.60	20.00	16.30	12.60	9.30	6.20	3.60
250	260	43.80	39.20	34.60	30.00	25.50	21.90	18.20	14.50	10.90	7.80	5.00
260	270	46.20	41.60	37.00	32.40	27.80	23.80	20.10	16.40	12.80	9.40	6.40
270	280	48.90	44.00	39.40	34.80	30.20	25.70	22.00	18.30	14.70	11.00	7.90
280	290	51.80	46.40	41.80	37.20	32.60	28.00	23.90	20.20	16.60	12.90	9.50
290	300	54.70	49.10	44.20	39.60	35.00	30.40	25.80	22.10	18.50	14.80	11.20
300	310	57.60	52.00	46.60	42.00	37.40	32.80	28.10	24.00	20.40	16.70	13.10
310	320	60.50	54.90	49.40	44.40	39.80	35.20	30.50	25.90	22.30	18.60	15.00
320	330	63.40	57.80	52.30	46.80	42.20	37.60	32.90	28.30	24.20	20.50	16.90
330	340	66.50	60.70	55.20	49.60	44.60	40.00	35.30	30.70	26.10	22.40	18.80
340	350	69.70	63.60	58.10	52.50	47.00	42.40	37.70	33.10	28.50	24.30	20.70
350	360	72.90	66.70	61.00	55.40	49.80	44.80	40.10	35.50	30.90	26.30	22.60
360	370	76.10	69.90	63.90	58.30	52.70	47.20	42.50	37.90	33.30	28.70	24.50
370	380	79.30	73.10	66.90	61.20	55.60	50.00	44.90	40.30	35.70	31.10	26.50
380	390	82.50	76.30	70.10	64.10	58.50	52.90	47.40	42.70	38.10	33.50	28.90
390	400	85.70	79.50	73.30	67.20	61.40	55.80	50.30	45.10	40.50	35.90	31.30
400	410	88.90	82.70	76.50	70.40	64.30	58.70	53.20	47.60	42.90	38.30	33.70
410	420	92.10	85.90	79.70	73.60	67.40	61.60	56.10	50.50	45.30	40.70	36.10

Merlin's Home Computers
Monthly Sales of Micros
19—

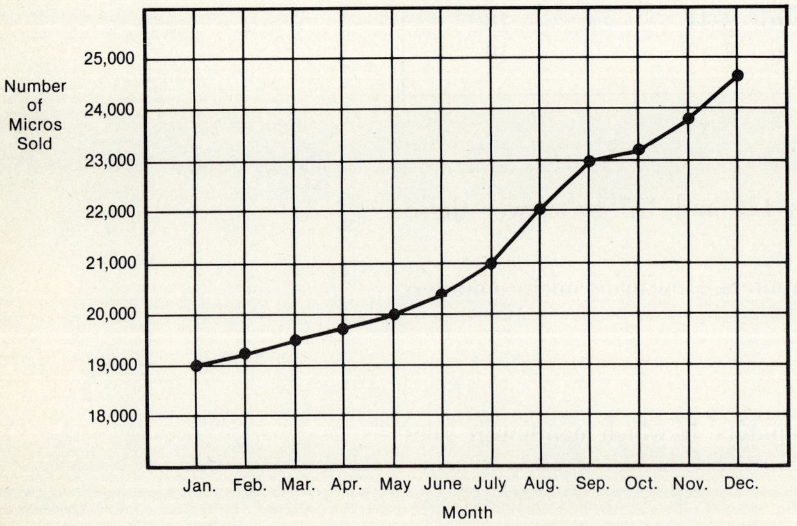

7. Refer to the graph of retail stores. What were the sales in dollars at automotive dealers?
a. 155 billion **b.** 145 billion
c. 140 billion **d.** 150 billion

8. Refer to the graph of retail stores. How much lower were sales at gasoline service stations than were sales at food stores?
a. 150 billion **b.** 55 billion
c. 95 billion **d.** 85 billion

ANSWER COLUMN

7. ______
8. ______
9. ______
10. ______

Retail Stores Sales by Kind of Business
Period Ending 19—

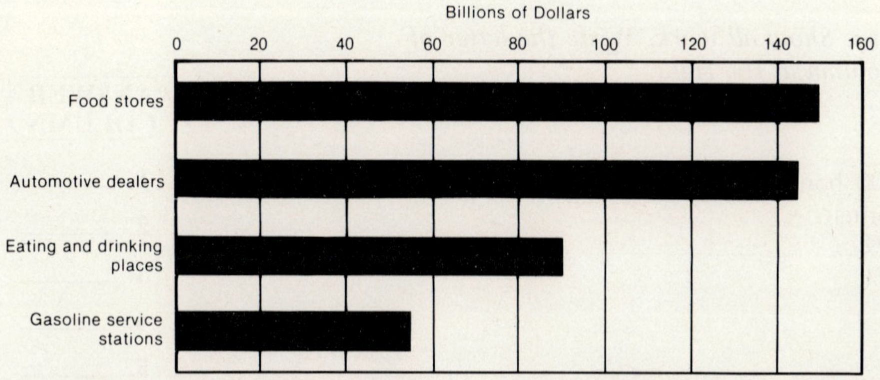

9. Refer to the sales graph of ABC Electronics Company. How many percent points higher were sales of portable TV sets than were sales of radios?
a. 5% **b.** 30%
c. 25% **d.** 20%

10. Refer to the sales graph of ABC Electronics Company. What percent of total sales were sales of portable and color TV sets?
a. 80% **b.** 45%
c. 25% **d.** 60%

Sales of ABC Electronics Company
Year Ending 19—

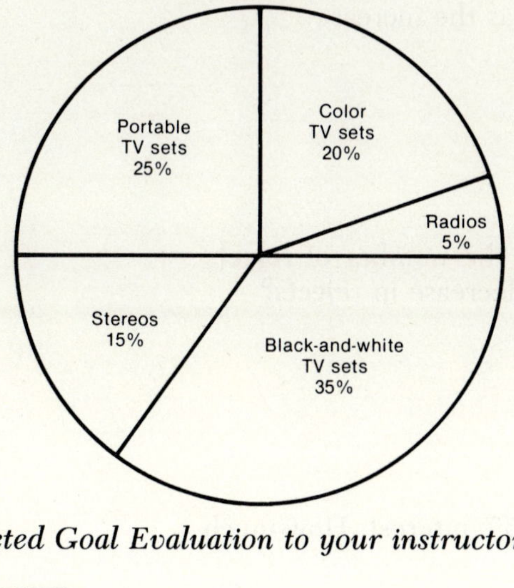

Give the completed Goal Evaluation to your instructor.

GOAL EVALUATION

Section 1 ____ %
2 ____ %
☐ NEXT GOAL
☐ Plan review

← WRITE YOUR SCORE ON YOUR TRACK RECORD

The checked box tells you what to do next.

NAME ....

DATE ....

GOAL EVALUATION

GOAL 18

USING EQUATIONS AND BUSINESS FORMULAS

Solve each of the following problems. Show all work. Write the letter of the correct answer in the answer column at the right.

DIRECTIONS. Solve.

ANSWER COLUMN

1. ______
2. ______
3. ______
4. ______
5. ______

1. Carl Himlee received an $1,800 bonus. His salary is $18,000. What percent of his salary was his bonus?
a. 110% **b.** 90%
c. 18% **d.** 10%

2. Rasa Typing Service paid 8% sales tax on a $4,000 word processor. How much tax was paid?
a. $320 **b.** $40
c. $4,320 **d.** $3,680

3. Potor Paper Products increased sales by 30% from last year to this year. If last year's sales were $400,000, what was the increase?
a. $60,000 **b.** $120,000
c. $280,000 **d.** $520,000

4. Kwik Type's new printing machine decreased the number of rejects per 1,000 from 4 to 3. What is the percent of decrease in rejects?
a. .1% **b.** 75%
c. 133% **d.** 25%

5. Florence Niely borrowed $5,000 for 1 year at 15% interest. How much interest did she pay?
a. $5,750 **b.** $5,250
c. $750 **d.** $150

Go on to the next page.

ANSWER COLUMN

6. ______
7. ______
8. ______
9. ______
10. ______

6. Lopez Catering Service received a 4% cash discount on a $15,000 order. What was the discount?

a. $15,600 b. $300
c. $600 d. $14,400

7. Dr. Maria Chang borrowed $3,600 for 1 year to buy a new X-ray machine. The interest on the loan was $540. What was the rate of interest?

a. $7\frac{1}{2}\%$ b. 15%
c. $6\frac{2}{3}\%$ d. $13\frac{1}{3}\%$

8. The production cost of a computer part is $8.19. If D&B Electronics marks up their products $66\frac{2}{3}\%$ based on the cost, what is the markup?

a. $10.92 b. $13.65
c. $2.73 d. $5.46

9. Davis Paper Products pays its sales representatives a 12% commission. If Lee Kosinski sold $12,000 in May, what was his commission?

a. $1,000 b. $10,560
c. $13,440 d. $1,440

10. Thomas Wilson gets a $200 salary plus a 4% commission on his sales. His pay was $344 last week. What were his sales?

a. $3,600 b. $1,376
c. $86 d. $8,600

Give the completed Goal Evaluation to your instructor.

GOAL EVALUATION

_____% ← WRITE YOUR SCORE ON YOUR TRACK RECORD

☐ NEXT GOAL
☐ Plan review

The checked box tells you what to do next.

NAME ....

DATE ....

GOAL EVALUATION

GOAL 19 **SOLVING WORD PROBLEMS**

Solve each of the following problems. Show all work. Write the letter of the correct answer in the answer column at the right.

DIRECTIONS. If there are not enough facts to answer a problem, choose the answer "more facts needed."

ANSWER COLUMN

1. ______
2. ______
3. ______

1. How many 5-gallon cans of oil can be filled from a 55-gallon drum?

a. more facts needed **b.** 11
c. 10 **d.** 50

2. Straus Paint had 94 gallons of enamel on hand when the spring sale started. At the end of the sale they had 32 gallons left. The price of the enamel was $9.00 a gallon. What was their profit?

a. more facts needed **b.** 62 gallons
c. $558 **d.** $846

3. Juan Felices compared this January's bills with those from last January for cleaning fluid. This year he used 75 pints at $1.47 a pint. Last year he used 97 pints at $1.29 a pint. How much less fluid did he use this year?

a. more facts needed **b.** $.18
c. 22 pints **d.** $14.88

Go on to the next page.

ANSWER COLUMN

4. ______

5. ______

6. ______

7. ______

4. Potobac Steel Products charges a $5 delivery fee on small orders. J&R Custom Fences received 6 orders last month costing a total of $947.72. What were the delivery charges?

a. more facts needed **b.** $30
c. $0 **d.** $977.72

5. It costs TLC Custom Service $1.29 to process an order. They get 7,200 orders every 3 months. What is the cost of processing these 7,200 orders?

a. more facts needed **b.** $3.87
c. $3,096 **d.** $9,288

6. Shopeaze Supermarket has a 40% markup based on cost on frozen foods. What is the markup on frozen pies that cost the supermarket $1.75 each?

a. more facts needed **b.** $.70
c. $2.45 **d.** $1.05

7. Last year Village Delivery Service budgeted $15,000 for insurance. This year they budgeted $12,500 for insurance, $10,000 for advertising, and $22,500 for new equipment. What was the percent of decrease in the amount budgeted for insurance?

a. more facts needed **b.** $16\frac{2}{3}$%
c. 120% **d.** $33\frac{1}{3}$%

Go on to the next page.

8. Elizabeth Fitz Henry worked 22 hours overtime last week. If overtime pay is $1\frac{1}{2}$ times the regular rate, how much was her overtime pay?
   a. more facts needed **b.** 62 hours
   c. 33 hours **d.** 73 hours

ANSWER COLUMN

8. ______

9. ______

10. ______

9. Salespeople at Space Age Computers whose sales are less than $2,500 a week get 4% commission on total sales. Salespeople whose sales are $2,500 or more get an extra 1% commission on total sales. One week Martin Edgerly had $4,100 of sales. What were his commissions?
   a. more facts needed **b.** $205
   c. $164 **d.** $180

10. The Golden Calf Restaurant found that 40% of its customers were tourists. If they serve 900 people a week, about how many were tourists?
   a. more facts needed **b.** 540
   c. 1,260 **d.** 360

Give the completed Goal Evaluation to your instructor.

GOAL EVALUATION

_____% ← WRITE YOUR SCORE ON YOUR TRACK RECORD

☐ NEXT GOAL
☐ Plan review

The checked box tells you what to do next.

NAME ....

DATE ....

GOAL EVALUATION

GOAL 20

RENAMING COMMON FRACTIONS TO SIMPLEST (LOWEST) TERMS

Solve each of the following problems. Show all work. Write the letter of the correct answer in the answer column at the right.

DIRECTIONS. Rename each fraction to lowest terms.

1. $\frac{42}{63}$

a. $\frac{4}{6}$
b. $\frac{14}{21}$
c. $\frac{2}{3}$
d. $\frac{6}{9}$

2. $\frac{48}{72}$

a. $\frac{12}{18}$
b. $\frac{8}{12}$
c. $\frac{4}{7}$
d. $\frac{2}{3}$

3. $\frac{54}{90}$

a. $\frac{9}{15}$
b. $\frac{3}{5}$
c. $\frac{5}{9}$
d. $\frac{27}{45}$

4. $\frac{24}{96}$

a. $\frac{2}{9}$
b. $\frac{3}{12}$
c. $\frac{1}{4}$
d. $\frac{4}{16}$

5. $\frac{36}{84}$

a. $\frac{3}{8}$
b. $\frac{6}{14}$
c. $\frac{9}{21}$
d. $\frac{3}{7}$

ANSWER COLUMN

1. ______
2. ______
3. ______
4. ______
5. ______

Go on to the next page.

6. $\frac{54}{66}$

a. $\frac{9}{11}$
b. $\frac{27}{33}$
c. $\frac{5}{6}$
d. $\frac{18}{22}$

7. $\frac{39}{78}$

a. $\frac{1}{2}$
b. $\frac{3}{7}$
c. $\frac{13}{26}$
d. $\frac{3}{6}$

ANSWER COLUMN

6. ______
7. ______
8. ______
9. ______
10. ______

8. $\frac{26}{65}$

a. $\frac{2}{6}$
b. $\frac{2}{5}$
c. $\frac{1}{3}$
d. $\frac{3}{8}$

9. $\frac{180}{240}$

a. $\frac{18}{24}$
b. $\frac{9}{12}$
c. $\frac{2}{3}$
d. $\frac{3}{4}$

10. $\frac{4,200}{7,000}$

a. $\frac{14}{35}$
b. $\frac{42}{70}$
c. $\frac{4}{7}$
d. $\frac{3}{5}$

Give the completed Goal Evaluation to your instructor.

GOAL EVALUATION

_____% ← WRITE YOUR SCORE ON YOUR TRACK RECORD

☐ NEXT GOAL
☐ Plan review

The checked box tells you what to do next.

NAME ____________

DATE ____________

GOAL EVALUATION

GOAL 21

RENAMING MIXED NUMBERS AS COMMON FRACTIONS

Solve each of the following problems. Show all work. Write the letter of the correct answer in the answer column at the right.

DIRECTIONS. Choose the number that makes each statement true.

1. $17 = \underline{\ ?\ }$

a. $2\frac{3}{7}$
b. $\frac{126}{7}$
c. $\frac{119}{7}$
d. $\frac{79}{7}$

2. $28\frac{2}{7} = \underline{\ ?\ }$

a. $\frac{30}{7}$
b. $\frac{282}{7}$
c. $\frac{56}{7}$
d. $\frac{198}{7}$

3. $47\frac{5}{6} = \underline{\ ?\ }$

a. $\frac{287}{6}$
b. $\frac{475}{6}$
c. $\frac{52}{6}$
d. $\frac{235}{6}$

4. $8\frac{4}{5} = \underline{\ ?\ }$

a. $\frac{84}{5}$
b. $\frac{12}{5}$
c. $\frac{32}{5}$
d. $\frac{44}{5}$

5. $27\frac{5}{8} = \underline{\ ?\ }$

a. $\frac{221}{8}$
b. $\frac{275}{8}$
c. $\frac{32}{8}$
d. $\frac{135}{8}$

ANSWER COLUMN

1. ______
2. ______
3. ______
4. ______
5. ______

Go on to the next page.

DIRECTIONS. Rename these improper fractions as mixed numbers.

ANSWER COLUMN

6. ______
7. ______
8. ______
9. ______
10. ______

6. $\frac{517}{9}$ = ?

a. $56\frac{1}{3}$
b. $51\frac{7}{9}$
c. $56\frac{7}{9}$
d. $57\frac{4}{9}$

7. $\frac{155}{7}$ = ?

a. $22\frac{1}{7}$
b. $23\frac{4}{7}$
c. $15\frac{5}{7}$
d. $20\frac{5}{7}$

8. $\frac{315}{8}$ = ?

a. $39\frac{5}{8}$
b. $38\frac{1}{8}$
c. $39\frac{3}{8}$
d. $31\frac{5}{8}$

9. $\frac{345}{6}$ = ?

a. $59\frac{1}{6}$
b. $57\frac{1}{2}$
c. $58\frac{1}{6}$
d. $34\frac{5}{6}$

10. $\frac{409}{11}$ = ?

a. $37\frac{7}{11}$
b. $37\frac{2}{11}$
c. $39\frac{7}{11}$
d. $40\frac{9}{11}$

Give the completed Goal Evaluation to your instructor.

GOAL EVALUATION

____% ← WRITE YOUR SCORE ON YOUR TRACK RECORD

☐ NEXT GOAL
☐ Plan review

The checked box tells you what to do next.

NAME ____________

DATE ____________

GOAL EVALUATION

GOAL 22 **MULTIPLYING COMMON FRACTIONS**

Solve each of the following problems. Show all work. Write the letter of the correct answer in the answer column at the right.

DIRECTIONS. Multiply.

ANSWER COLUMN

1. ______
2. ______
3. ______
4. ______
5. ______

1. $\frac{2}{5} \times \frac{1}{4}$

- **a.** $1\frac{3}{5}$
- **b.** $\frac{2}{20}$ or $\frac{1}{10}$
- **c.** $\frac{3}{9}$ or $\frac{1}{3}$
- **d.** 1

2. $\$60 \times 3\frac{1}{3}$

- **a.** \$20
- **b.** $\$180\frac{1}{3}$
- **c.** \$18
- **d.** \$200

3. $\frac{3}{10} \times \frac{1}{4}$

- **a.** $\frac{4}{14}$ or $\frac{2}{7}$
- **b.** $1\frac{1}{5}$
- **c.** $\frac{3}{40}$
- **d.** $\frac{10}{12}$ or $\frac{5}{6}$

4. $88 \times 2\frac{1}{2}$

- **a.** $33\frac{1}{5}$
- **b.** $176\frac{1}{2}$
- **c.** 220
- **d.** 44

5. $\frac{3}{5} \times \frac{3}{4}$

- **a.** $\frac{2}{3}$
- **b.** $1\frac{1}{4}$
- **c.** $\frac{4}{5}$
- **d.** $\frac{9}{20}$

Give the completed Goal Evaluation to your instructor.

GOAL EVALUATION
____% ← WRITE YOUR SCORE ON YOUR TRACK RECORD ☐ NEXT GOAL ☐ Plan review

The checked box tells you what to do next.

NAME ....

DATE ....

GOAL EVALUATION

GOAL 23 **DIVIDING WITH COMMON FRACTIONS**

Solve each of the following problems. Show all work. Write the letter of the correct answer in the answer column at the right.

DIRECTIONS. Divide.

ANSWER COLUMN

1. ______
2. ______
3. ______
4. ______
5. ______

1. 2,817 ÷ $\frac{9}{10}$

a. 2,535$\frac{3}{10}$

b. 281$\frac{7}{10}$

c. 3,130

d. 313

2. $72.80 ÷ $\frac{4}{5}$

a. $58.24

b. $91

c. $3.64

d. $14.56

3. 1,245 ÷ $\frac{3}{8}$

a. 51$\frac{7}{8}$

b. 155$\frac{5}{8}$

c. 466$\frac{7}{8}$

d. 3,320

4. 4,120 ÷ $\frac{2}{3}$

a. 2,746$\frac{2}{3}$

b. 6,180

c. 1,373$\frac{1}{3}$

d. 686$\frac{2}{3}$

5. $24.53 ÷ 2$\frac{1}{5}$

a. $12.27

b. $4.91

c. $11.15

d. $53.97

Go on to the next page.

ANSWER COLUMN

6. ______
7. ______
8. ______
9. ______
10. ______

6. $3{,}618 \div 1\frac{1}{8}$

a. 3,216
b. $4{,}070\frac{1}{4}$
c. $452\frac{1}{4}$
d. $50\frac{1}{4}$

7. $1{,}179 \div 13\frac{1}{10}$

a. $117\frac{9}{10}$
b. $15{,}444\frac{9}{10}$
c. $900\frac{1}{10}$
d. 90

8. $16.25 ÷ $6\frac{1}{4}$

a. $2.60
b. $4.06
c. $65
d. $101.56

9. $1{,}492 \div 2\frac{2}{3}$

a. $3{,}978\frac{2}{3}$
b. $994\frac{2}{3}$
c. $559\frac{1}{2}$
d. $746\frac{2}{3}$

10. $18.75 ÷ $\frac{5}{8}$

a. $.47
b. $30
c. $11.72
d. $2.35

Give the completed Goal Evaluation to your instructor.

GOAL EVALUATION

_____% ← WRITE YOUR SCORE ON YOUR TRACK RECORD

☐ NEXT GOAL
☐ Plan review

The checked box tells you what to do next.

NAME ....

DATE ....

GOAL EVALUATION

GOAL 24 ADDING COMMON FRACTIONS

Solve each of the following problems. Show all work. Write the letter of the correct answer in the answer column at the right.

DIRECTIONS. Add.

1. $\frac{3}{5} + \frac{7}{5} + \frac{4}{5} + \frac{2}{5}$

a. $1\frac{3}{5}$
b. $3\frac{1}{5}$
c. $\frac{11}{25}$
d. $\frac{4}{5}$

2. $\frac{8}{10} + \frac{5}{10} + \frac{3}{10} + \frac{17}{10}$

a. $3\frac{3}{10}$
b. $\frac{33}{40}$
c. $1\frac{13}{20}$
d. $\frac{33}{100}$

3. $\frac{9}{4} + \frac{3}{8} + \frac{7}{8} + \frac{3}{4}$

a. $1\frac{5}{6}$
b. $\frac{11}{12}$
c. $\frac{11}{16}$
d. $4\frac{1}{4}$

4. $\frac{3}{2} + \frac{4}{5} + \frac{3}{4} + \frac{3}{10}$

a. $\frac{19}{21}$
b. $\frac{19}{20}$
c. $3\frac{7}{20}$
d. $3\frac{4}{5}$

5. $\frac{4}{3} + \frac{4}{5} + \frac{3}{4} + \frac{5}{2}$

a. $5\frac{23}{60}$
b. $1\frac{1}{7}$
c. $\frac{4}{15}$
d. 2

ANSWER COLUMN

1. ______
2. ______
3. ______
4. ______
5. ______

Give the completed Goal Evaluation to your instructor.

GOAL EVALUATION

_____% ← WRITE YOUR SCORE ON YOUR TRACK RECORD

☐ NEXT GOAL
☐ Plan review

The checked box tells you what to do next.

NAME ....

DATE ....

GOAL EVALUATION

GOAL 25

SUBTRACTING COMMON FRACTIONS

Solve each of the following problems. Show all work. Write the letter of the correct answer in the answer column at the right.

DIRECTIONS. Subtract.

1. $\frac{5}{6} - \frac{2}{3}$

- **a.** $\frac{3}{3}$ or 1
- **b.** $\frac{3}{6}$ or $\frac{1}{2}$
- **c.** $\frac{3}{9}$ or $\frac{1}{3}$
- **d.** $\frac{3}{18}$ or $\frac{1}{6}$

2. $\frac{7}{10} - \frac{1}{3}$

- **a.** $\frac{11}{30}$
- **b.** $\frac{11}{7}$ or $1\frac{4}{7}$
- **c.** $\frac{6}{7}$
- **d.** $\frac{6}{30}$ or $\frac{1}{5}$

3. $\frac{7}{8} - \frac{3}{4}$

- **a.** $\frac{4}{4}$ or 1
- **b.** $\frac{4}{12}$ or $\frac{1}{3}$
- **c.** $\frac{4}{32}$ or $\frac{1}{8}$
- **d.** $\frac{4}{8}$ or $\frac{1}{2}$

4. $\frac{5}{10} - \frac{2}{5}$

- **a.** $\frac{3}{5}$
- **b.** $\frac{3}{50}$
- **c.** $\frac{1}{10}$
- **d.** $\frac{3}{10}$

5. $\frac{3}{5} - \frac{1}{2}$

- **a.** $\frac{1}{3}$
- **b.** $\frac{1}{10}$
- **c.** $\frac{2}{3}$
- **d.** $\frac{2}{10}$ or $\frac{1}{5}$

ANSWER COLUMN

1. ______
2. ______
3. ______
4. ______
5. ______

Go on to the next page.

6. $8\frac{7}{10} - 1\frac{4}{5}$
   a. $6\frac{45}{50}$ or $6\frac{9}{10}$
   b. $7\frac{3}{10}$
   c. $7\frac{3}{5}$
   d. $7\frac{45}{50}$ or $7\frac{9}{10}$

7. $5\frac{1}{2} - 2\frac{5}{8}$
   a. $3\frac{4}{6}$ or $3\frac{2}{3}$
   b. $2\frac{4}{8}$ or $2\frac{1}{2}$
   c. $2\frac{7}{8}$
   d. $3\frac{7}{8}$

ANSWER COLUMN

6. ______
7. ______
8. ______
9. ______
10. ______

8. $7\frac{5}{8} - 3\frac{3}{5}$
   a. $4\frac{2}{40}$ or $4\frac{1}{20}$
   b. $4\frac{1}{40}$
   c. $4\frac{2}{3}$
   d. $4\frac{1}{3}$

9. $17\frac{3}{8} - 12\frac{1}{2}$
   a. $5\frac{2}{6}$ or $5\frac{1}{3}$
   b. $5\frac{2}{8}$ or $5\frac{1}{4}$
   c. $5\frac{7}{8}$
   d. $4\frac{7}{8}$

10. $9\frac{3}{10} - 4\frac{1}{2}$
   a. $5\frac{2}{8}$ or $5\frac{1}{4}$
   b. $5\frac{2}{10}$ or $5\frac{1}{5}$
   c. $4\frac{8}{10}$ or $4\frac{4}{5}$
   d. $5\frac{8}{10}$ or $5\frac{4}{5}$

Give the completed Goal Evaluation to your instructor.

GOAL EVALUATION

____% ← WRITE YOUR SCORE ON YOUR TRACK RECORD

☐ NEXT GOAL
☐ Plan review

The checked box tells you what to do next.

WRITE YOUR SCORE ON YOUR TRACK RECORD

NAME ....

DATE ....

ALTERNATE GOAL EVALUATION

GOAL 1 READING AND WRITING NUMBERS

Solve each of the following problems. Write the letter of the correct answer in the answer column at the right.

DIRECTIONS. Choose the answer that shows how each number should be read.

1. 6,007,407
a. six million, seven hundred thousand, four hundred and seven
b. six million, seventy-four hundred and seven
c. six million, seventy thousand, four hundred and seven
d. six million, seven thousand, four hundred seven

2. 96.412
a. ninety-six point four one two
b. ninety-six and four hundred twelve ten-thousandths
c. ninety-six point four hundred twelve thousandths
d. ninety-six four hundred twelve thousandths

3. 706,412
a. seventy-six thousand four hundred twelve
b. seven hundred six thousand four hundred twelve
c. seven hundred six thousand, four hundred twelve
d. seven hundred and six thousand, four hundred and twelve

4. .013
a. thirteen hundredths
b. thirteen ten thousandths
c. point thirteen thousandths
d. thirteen thousandths

DIRECTIONS. Choose the number that shows how each number should be written in digits.

5. Seven million, one hundred sixty thousand
a. 7,100,060 **b.** 7,000,160
c. 7,160,000 **d.** 7,016,000

6. Sixty-four thousand and seventeen ten thousandths
a. 64,017 **b.** 64,000.0017
c. 64,000.17 **d.** 64,000.017

7. Eighty-five and seven hundredths
a. 85.07 **b.** .857
c. .8507 **d.** 85.007

ANSWER COLUMN

1. ______
2. ______
3. ______
4. ______
5. ______
6. ______
7. ______

Go on to the next page.

8. Sixty million twelve

a. 60,012 **b.** 60,000,012

c. 60,000,000.12 **d.** 60,012,000

9. Seven thousand, two hundred fifty-four

a. 7,000,254 **b.** 720,054

c. 7,200,054 **d.** 7,254

10. Sixty-two ten thousandths

a. 62.0001 **b.** .062

c. .0062 **d.** .62

ANSWER COLUMN

8. ______

9. ______

10. ______

Give the completed Alternate Goal Evaluation to your instructor.

____% ← WRITE YOUR SCORE ON YOUR TRACK RECORD
☐ NEXT GOAL
☐ Plan review

The checked box tells you what to do next.

NAME ....

DATE ....

ALTERNATE GOAL EVALUATION

GOAL 2

ROUNDING OFF NUMBERS AND MONEY AMOUNTS

Solve each of the following problems. Write the letter of the correct answer in the answer column at the right.

DIRECTIONS. Choose the answer that shows how each number should be rounded.

1. Round off 41 to the nearest ten.
a. 40 **b.** 50 **c.** 41 **d.** 30

2. Round off 7,453.2 to the nearest hundred.
a. 7,400 **b.** 7,000 **c.** 7,500 **d.** 7,450

3. Round off 12,401 to the nearest thousand.
a. 13,000 **b.** 12,000 **c.** 12,400 **d.** 12,500

4. Round off $3.75 to the nearest ten dollars.
a. $0 **b.** $3.80 **c.** $10 **d.** $4.00

5. Round off 149,713 to the nearest hundred thousand.
a. 200,000 **b.** 150,000 **c.** 149,000 **d.** 100,000

6. Round off 49.2 to the nearest hundred.
a. 49 **b.** 100 **c.** 0 **d.** 50

7. Round off 72.489 to the nearest tenth.
a. 70 **b.** 72 **c.** 72.49 **d.** 72.5

8. Round off $254.3618 to the nearest tenth of a cent.
a. $254.362 **b.** $254.40 **c.** $254.37 **d.** $254.36

9. Round down $199.64 to the nearest dollar.
a. $200 **b.** $190 **c.** $199 **d.** $191

10. Round up $.4548 to the nearest cent.
a. $.46 **b.** $.50 **c.** $.45 **d.** $.455

ANSWER COLUMN

1. ______
2. ______
3. ______
4. ______
5. ______
6. ______
7. ______
8. ______
9. ______
10. ______

Give the completed Goal Evaluation to your instructor.

_____% ← WRITE YOUR SCORE ON YOUR TRACK RECORD

☐ NEXT GOAL
☐ Plan review

The checked box tells you what to do next.

NAME ....

DATE ....

ALTERNATE GOAL EVALUATION

GOAL 3

ADDING WHOLE NUMBERS AND MONEY AMOUNTS

Solve each of the following problems. Show all work. Write the letter of the correct answer in the answer column at the right.

•SECTION 1

DIRECTIONS. Add.

ANSWER COLUMN

1. ______
2. ______
3. ______
4. ______
5. ______
6. ______

1. 32 + 51 + 82 + 69

a. 218 **b.** 224 **c.** 234 **d.** 2,214

2. 42 + 86 + 41 + 25

a. 122 **b.** 184 **c.** 194 **d.** 1,814

3. 25 + 69 + 77 + 47

a. 137 **b.** 1,928 **c.** 198 **d.** 218

4. 59 + 67 + 98 + 74

a. 298 **b.** 278 **c.** 2,728 **d.** 235

5. 62 + 33 + 35 + 56

a. 176 **b.** 186 **c.** 1,716 **d.** 123

•SECTION 2

DIRECTIONS. Add.

6. \$30,665 + 72,396 + 835.57 + 779.85

a. \$104,675.142 **b.** \$102,456.35 **c.** \$104,676.42 **d.** \$104,575.42

Go on to the next page.

ANSWER COLUMN

7. ______
8. ______
9. ______
10. ______

7.

$49,756	**a.** $79,530.51
715.74	**b.** $81,650.161
531.87	**c.** $81,641.16
+30,648	**d.** $81,651.61

8. $82,321 + $346.64 + $689.76 + $45,834
- **a.** $129,191.40
- **b.** $127,070.30
- **c.** $129,190.140
- **d.** $231,795

9. $478.55 + $78,395 + $.68 + $938.88
- **a.** $78,691.91
- **b.** $79,811.211
- **c.** $221,206
- **d.** $79,813.11

10. $353.69 + $54,213 + $689.73 + $48,745
- **a.** $207,300
- **b.** $104,001.42
- **c.** $92,880.32
- **d.** $104,000.142

Give the completed Goal Evaluation to your instructor.

Section 1______%
2______%
☐ NEXT GOAL
☐ Plan review

← WRITE YOUR SCORE ON YOUR TRACK RECORD

The checked box tells you what to do next.

NAME ....

DATE ....

ALTERNATE GOAL EVALUATION

GOAL 4 SUBTRACTING WHOLE NUMBERS AND MONEY AMOUNTS

Solve each of the following problems. Show all work. Write the letter of the correct answer in the answer column at the right.

•SECTION 1

DIRECTIONS. Subtract.

1. $46 - 27$

a. 29 **b.** 21 **c.** 19 **d.** 20

2. $70 - 23$

a. 47 **b.** 53 **c.** 57 **d.** 50

3. $85 - 57$

a. 30 **b.** 38 **c.** 32 **d.** 28

4. $74 - 48$

a. 30 **b.** 34 **c.** 26 **d.** 36

5. $88 - 29$

a. 61 **b.** 59 **c.** 69 **d.** 60

•SECTION 2

DIRECTIONS. Subtract.

6. $64{,}738 - 35{,}809$

a. 28,909 **b.** 28,929 **c.** 31,131 **d.** 39,939

7. $57{,}008 - 18{,}263$

a. 19,843 **b.** 38,745 **c.** 41,265 **d.** 49,845

ANSWER COLUMN

1. ______
2. ______
3. ______
4. ______
5. ______
6. ______
7. ______

Go on to the next page.

8. $853.60 − $375.27
a. $288.43
b. $588.43
c. $478.33
d. $522.47

9. $69,416 − $206.18
a. $69,200.82
b. $69,209.82
c. $69,210.92
d. $48,798

ANSWER COLUMN

8. ______

9. ______

10. ______

10. $754.36 − $180.73
a. $573.63
b. $570.63
c. $634.43
d. $674.63

Give the completed Goal Evaluation to your instructor.

Section 1 ______%	
2 ______%	← WRITE YOUR SCORE ON YOUR TRACK RECORD
☐ NEXT GOAL	
☐ Plan review	

The checked box tells you what to do next.

NAME ....

DATE ....

ALTERNATE GOAL EVALUATION

GOAL 5

MULTIPLYING WHOLE NUMBERS AND MONEY AMOUNTS

Solve each of the following problems. Show all work. Write the letter of the correct answer in the answer column at the right.

•SECTION 1

DIRECTIONS. Multiply.

1. 357×9
   a. 2,753
   b. 3,213
   c. 3,693
   d. 274,563

2. 913×6
   a. 5,468
   b. 6,028
   c. 5,478
   d. 54,618

3. 146×6
   a. 646
   b. 62,436
   c. 3,026
   d. 876

4. 704×22
   a. 1,628
   b. 2,816
   c. 15,488
   d. 14,488

5. 324×200
   a. 648
   b. 60,048
   c. 64,800
   d. 60,408

ANSWER COLUMN

1. ______
2. ______
3. ______
4. ______
5. ______

Go on to the next page.

•SECTION 2

ANSWER COLUMN

6. ______
7. ______
8. ______
9. ______
10. ______

DIRECTIONS. Multiply.

6. $84.27 × 57

a. $4,803.39
b. $1,011.24
c. $5,017.39
d. $4,804.39

7. $28.17 × 37

a. $281.10
b. $1,042.29
c. $1,354.29
d. $1,556.29

8. 8,943 × 765

a. 160,974
b. 7,825,495
c. 6,841,395
d. 6,630,395

9. 6,547 × 209

a. 72,017
b. 189,863
c. 1,368,323
d. 1,703,323

10. 5,924 × 418

a. 77,012
b. 2,939,532
c. 3,189,232
d. 2,476,232

Give the completed Goal Evaluation to your instructor.

Section 1______%
2______%
☐ NEXT GOAL
☐ Plan review

← WRITE YOUR SCORE ON YOUR TRACK RECORD

The checked box tells you what to do next.

NAME ....

DATE ....

ALTERNATE GOAL EVALUATION

GOAL 6 **DIVIDING WHOLE NUMBERS AND MONEY AMOUNTS**

Solve each of the following problems. Show all work. Write the letter of the correct answer in the answer column at the right.

•SECTION 1

DIRECTIONS. Divide and show what the remainder is, if any.

1. $8\overline{)6{,}507}$

a. 82 R1
b. 813 R3
c. 821
d. 820 R1

2. $6\overline{)\$56.35}$

a. $11.01
b. $93.91
c. $9.30 R55
d. $9.39 R1

3. $3\overline{)6{,}623}$

a. 2,207 R2
b. 227 R2
c. 22,072
d. 2,272

4. $9\overline{)\$86.09}$

a. $96.55
b. $9.50 R29
c. $9.56 R5
d. $9.56

5. $6\overline{)6{,}049}$

a. 108 R1
b. 1,800 R1
c. 1,008 R1
d. 18 R1

ANSWER COLUMN

1. ______
2. ______
3. ______
4. ______
5. ______

Go on to the next page.

•SECTION 2

ANSWER COLUMN

6. ______

7. ______

8. ______

9. ______

10. ______

6. $136\overline{)95{,}989}$

a. 750 R109
b. 705 R209
c. 75 R109
d. 705 R109

7. $395\overline{)\$2{,}042.94}$

a. $5.17 R79
b. $517.79
c. $5.17 R2
d. $517 R79

8. $212\overline{)\$1{,}927.61}$

a. $909.53
b. $9.09 R53
c. $9.90 R53
d. $9.9 R53

9. $762\overline{)384{,}239}$

a. 54 R191
b. 50 R539
c. 504 R191
d. 540 R191

10. $84\overline{)92{,}821}$

a. 1,151
b. 115 R1
c. 1,105 R1
d. 1,150 R1

Give the completed Goal Evaluation to your instructor.

Section 1______%
2______%
☐ NEXT GOAL
☐ Plan review

← WRITE YOUR SCORE ON YOUR TRACK RECORD

The checked box tells you what to do next.

NAME ....

DATE ....

ALTERNATE GOAL EVALUATION

GOAL 7

PLACING THE DECIMAL POINT

Solve each of the following problems. Show all work. Write the letter of the correct answer in the answer column at the right.

•SECTION 1

DIRECTIONS. Rewrite each problem and solve.

ANSWER COLUMN

1. ______
2. ______
3. ______
4. ______
5. ______

1. 9.447 + 7.3 + 6 + .7
 a. 22.1447
 b. 23.447
 c. 75.23
 d. 2.3447

2. 3.762 + 2.2 + 6 + 1.77
 a. 13.732
 b. 12.1732
 c. 1.3732
 d. 3.967

3. .08 + 7.63 + 3.3 + 2.023
 a. 1.3033
 b. 12.1033
 c. 13.033
 d. 28.27

4. 254 − 3.7
 a. 250.3
 b. 217
 c. 251.7
 d. 261.3

5. 656 − 45.7
 a. 611.7
 b. 610.3
 c. 199
 d. 61.03

Go on to the next page.

•SECTION 2

ANSWER COLUMN

6. ______

7. ______

8. ______

9. ______

10. ______

DIRECTIONS. Rewrite each problem and solve.

6. 33.4 × .582
- **a.** 33.2328
- **b.** 19.4388
- **c.** 35.328
- **d.** 1.94388

7. .006 × .05
- **a.** .03
- **b.** .003
- **c.** .0003
- **d.** .00003

8. 7.46 × 8.5
- **a.** 6.341
- **b.** 56.23
- **c.** 58.3
- **d.** 63.41

9. 27.05 ÷ 2.5
- **a.** 10.82
- **b.** 1.082
- **c.** 108.2
- **d.** 10.85

10. 560.24 ÷ .8
- **a.** 70.03
- **b.** 70.3
- **c.** 700.3
- **d.** 7,003

Give the completed Goal Evaluation to your instructor.

Section 1_____%
2_____%
☐ NEXT GOAL
☐ Plan review

← WRITE YOUR SCORE ON YOUR TRACK RECORD

The checked box tells you what to do next.

NAME ....

DATE ....

ALTERNATE GOAL EVALUATION

GOAL 8

WRITING COMMON FRACTIONS, DECIMAL FRACTIONS, AND PERCENTS

Solve each of the following problems. Show all work. Write the letter of the correct answer in the answer column at the right.

DIRECTIONS. Choose the number that makes each statement true.

1. __?__ of the diagram is shaded.
   a. $\frac{3}{5}$ b. $\frac{5}{8}$
   c. $\frac{8}{5}$ d. $\frac{5}{3}$

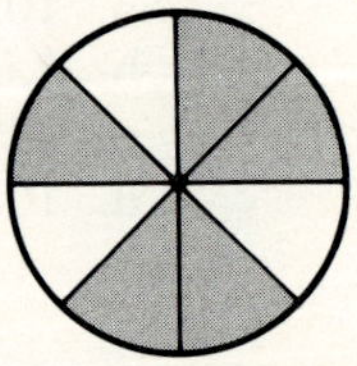

2. Olofson Bros. made a \$7 million profit last year and a \$9 million profit this year. This year's profit was __?__ of last year's profit.
   a. $\frac{7}{9}$ b. $\frac{9}{7}$
   c. $\frac{2}{9}$ d. 217

3. The change in the machine's design allows workers in CPM, Inc. to assemble 7 units for every 4 units assembled before. The number of units being assembled now as compared to before is __?__.
   a. $\frac{7}{4}$ b. $\frac{11}{4}$
   c. $\frac{4}{7}$ d. $\frac{11}{7}$

4. The decimal fraction .97 means 97 out of __?__.
   a. 100 b. 1,000
   c. 10 d. 1

5. The decimal fraction .04 means the same thing as the common fraction __?__.
   a. $\frac{4}{100}$ b. $\frac{4}{10}$
   c. $\frac{4}{1}$ d. $\frac{1}{4}$

ANSWER COLUMN

1. ______
2. ______
3. ______
4. ______
5. ______

Go on to the next page.

ANSWER COLUMN

6. ______
7. ______
8. ______
9. ______
10. ______

6. __?__ of the diagram is shaded.

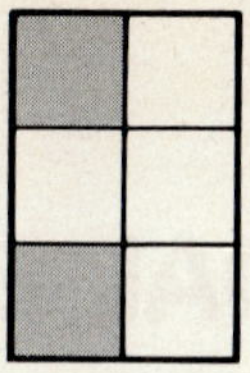

a. $\frac{2}{8}$ **b.** $\frac{6}{8}$
c. $\frac{2}{6}$ **d.** $\frac{6}{2}$

7. The decimal fraction .146 means the same thing as the common fraction __?__.

a. $\frac{1.46}{100}$ **b.** $\frac{146}{100}$
c. $\frac{146}{1,000}$ **d.** $\frac{146}{10,000}$

8. 75% means __?__ parts in every 100.

a. 7.5 **b.** .75
c. 750 **d.** 75

9. If 100% of your $750 investment was lost, __?__ is left?

a. $0 **b.** $740
c. $650 **d.** $74

10. Taxes were 50% of profits. The base of comparison in this statement is __?__.

a. taxes **b.** profits + taxes
c. profits **d.** profits − taxes

Give the completed Goal Evaluation to your instructor.

_____%
☐ NEXT GOAL
☐ Plan review

← WRITE YOUR SCORE ON YOUR TRACK RECORD

The checked box tells you what to do next.

NAME ....

DATE ....

ALTERNATE GOAL EVALUATION

GOAL 9

RENAMING COMMON FRACTIONS, DECIMAL FRACTIONS, AND PERCENTS

Solve each of the following problems. Show all work. Write the letter of the correct answer in the answer column at the right.

DIRECTIONS. Choose the number that makes the statement true.

1. $\frac{1}{4}$ = ___?___
- a. .14
- b. .25
- c. .4
- d. 1.4

2. $\frac{3}{5}$ = ___?___
- a. .35
- b. .167
- c. 6
- d. .6

3. $14\frac{6}{10}\%$ = ___?___
- a. .146
- b. 14.6
- c. 1.46
- d. .0146

4. 94% = ___?___
- a. 94
- b. .94
- c. 9.4
- d. .094

5. .197 = ___?___
- a. .197%
- b. 197%
- c. 19.7%
- d. 1.97%

6. 1.46% = ___?___
- a. 1.46
- b. .0146
- c. .146
- d. .00146

7. $9\frac{1}{4}$ = ___?___
- a. 9.14
- b. 9.4
- c. 9.41
- d. 9.25

8. $6\frac{3}{5}$ = ___?___
- a. 6.35
- b. 6.167
- c. 6.53
- d. 6.6

9. 5.361 = ___?___
- a. 53.61%
- b. 5.361%
- c. 536.1%
- d. .5361%

10. 9.97% = ___?___
- a. .0997
- b. 9.97
- c. .00997
- d. .997

ANSWER COLUMN

1. ______
2. ______
3. ______
4. ______
5. ______
6. ______
7. ______
8. ______
9. ______
10. ______

Give the completed Goal Evaluation to your instructor.

______% ← WRITE YOUR SCORE ON YOUR TRACK RECORD
□ NEXT GOAL
□ Plan review

The checked box tells you what to do next.

NAME ....
DATE ....

ALTERNATE GOAL EVALUATION

GOAL 10 **WORKING WITH TIME**

Solve each of the following problems. Show all work. Write the letter of the correct answer in the answer column at the right.

•SECTION 1

DIRECTIONS. Choose the number that makes each statement true.

1. 9 hr 15 min = ___?___
- **a.** 915 min
- **b.** 555 min
- **c.** 456 min
- **d.** 1,235 min

2. 146 min = ___?___
- **a.** 6 hr 6 min
- **b.** 2 hr 46 min
- **c.** 2 hr 26 min
- **d.** 1 hr 46 min

3. 31 hr 13 min − 17 hr 55 min = ___?___
- **a.** 13 hr 58 min
- **b.** 14 hr 18 min
- **c.** 13 hr 18 min
- **d.** 12 hr 18 min

4. 32 hr 11 min − 13 hr 57 min = ___?___
- **a.** 18 hr 54 min
- **b.** 19 hr 14 min
- **c.** 18 hr 14 min
- **d.** 17 hr 14 min

5. 11 hr 27 min + 17 hr + 32 hr 26 min + 42 min = ___?___
- **a.** 44 hr 52 min
- **b.** 61 hr 95 min
- **c.** 61 hr 35 min
- **d.** 62 hr 35 min

ANSWER COLUMN

1. ______
2. ______
3. ______
4. ______
5. ______
6. ______
7. ______

•SECTION 2

DIRECTIONS. Find the correct time or date.

6. How many days are there from September 13 to December 22?
- **a.** 102 days
- **b.** 99 days
- **c.** 101 days
- **d.** 100 days

7. What is the date 41 days after September 17?
- **a.** October 27
- **b.** October 28
- **c.** October 26
- **d.** October 25

Go on to the next page.

ANSWER COLUMN

8. ______
9. ______
10. ______

8. What is the date 75 days after February 2, 1996?
- **a.** April 18
- **b.** April 16
- **c.** April 19
- **d.** April 17

9. How many days are there from September 24 to October 10?
- **a.** 15 days
- **b.** 18 days
- **c.** 17 days
- **d.** 16 days

10. What is the date 9 weeks after September 27?
- **a.** November 28
- **b.** November 29
- **c.** November 27
- **d.** November 30

Give the completed Goal Evaluation to your instructor.

Section 1_____%
2_____%
☐ NEXT GOAL
☐ Plan review

← WRITE YOUR SCORE ON YOUR TRACK RECORD

The checked box tells you what to do next.

NAME ....

DATE ....

ALTERNATE GOAL EVALUATION

GOAL 11

IDENTIFYING U.S. CUSTOMARY UNITS

Solve each of the following problems. Show all work. Write the letter of the correct answer in the answer column at the right.

DIRECTIONS. Match the term in Column I with what they mean or measure in Column II.

Column I	Column II
1. foot	a. pound
2. 36 inches	b. weight
3. gallon	c. quart
4. 2 pints	d. temperature
5. degree	e. length
	f. yard
	g. capacity or volume

DIRECTIONS. Choose the unit that is closest in size to the object named.

6. The area of a parking lot
   a. 200 square miles b. 200 square yards
   c. 200 yards d. 200 square feet

7. The length of a bathtub
   a. $1\frac{1}{2}$ feet b. $1\frac{1}{2}$ square yards
   c. $1\frac{1}{2}$ yards d. $1\frac{1}{2}$ square feet

8. The quantity of soup needed to fill a large kettle
   a. 2 fluid ounces b. 2 pints
   c. 2 quarts d. 2 gallons

9. The weight of a 2-quart container of juice
   a. 4 pints b. 4 ounces
   c. 4 tons d. 4 pounds

10. The height of a can of soda
    a. 5 feet b. 5 miles
    c. 5 yards d. 5 inches

ANSWER COLUMN

1. ______
2. ______
3. ______
4. ______
5. ______
6. ______
7. ______
8. ______
9. ______
10. ______

Give the completed Goal Evaluation to your instructor.

_____% ← WRITE YOUR SCORE ON YOUR TRACK RECORD
☐ NEXT GOAL
☐ Plan review

The checked box tells you what to do next.

NAME ....

DATE ....

ALTERNATE GOAL EVALUATION

GOAL 12

RENAMING U.S. CUSTOMARY UNITS

Solve each of the following problems. Show all work. Write the letter of the correct answer in the answer column at the right.

DIRECTIONS. Choose the number that makes each statement true.

1. 256 gallons = __?__ quarts
- **a.** 16
- **b.** 1,024
- **c.** 8
- **d.** 4,096

2. 32,000 pounds = __?__ tons
- **a.** 16
- **b.** $888\frac{8}{9}$
- **c.** 32
- **d.** $2{,}666\frac{2}{3}$

3. 13 pounds = __?__ ounces
- **a.** 156
- **b.** 468
- **c.** 416
- **d.** 208

4. 105 yards = __?__ feet
- **a.** 1,260
- **b.** 1,680
- **c.** 315
- **d.** 35

5. 240 fluid ounces = __?__ pints
- **a.** 20
- **b.** $6\frac{2}{3}$
- **c.** 705
- **d.** 15

6. 56 pints = __?__ quarts
- **a.** $3\frac{1}{2}$
- **b.** 28
- **c.** $4\frac{2}{3}$
- **d.** $1\frac{3}{4}$

7. 72 square feet = __?__ square inches
- **a.** 6
- **b.** 2
- **c.** 10,368
- **d.** 864

8. 60,000 ounces = __?__ pounds
- **a.** 3,750
- **b.** 1,875
- **c.** 5,000
- **d.** $1{,}666\frac{2}{3}$

ANSWER COLUMN

1. ______

2. ______

3. ______

4. ______

5. ______

6. ______

7. ______

8. ______

Go on to the next page.

9. 2,400 square inches = ___?___ square feet

a. $16\frac{2}{3}$

b. $5\frac{1}{3}$

c. 28,800

d. 200

10. 270 square feet = ___?___ square yards

a. $22\frac{2}{3}$

b. 90

c. 30

d. $7\frac{1}{2}$

ANSWER COLUMN

9. ______

10. ______

Give the completed Goal Evaluation to your instructor.

____% ← WRITE YOUR SCORE ON YOUR TRACK RECORD
☐ NEXT GOAL
☐ Plan review

The checked box tells you what to do next.

NAME ....

DATE ....

ALTERNATE GOAL EVALUATION

GOAL 13 IDENTIFYING SI UNITS

Solve each of the following problems. Show all work. Write the letter of the correct answer in the answer column at the right.

•SECTION 1

DIRECTIONS. Match the terms in Column I with what they mean or measure in Column II.

Column I	Column II
1. deka	**a.** milligram
2. cg	**b.** 100
3. hecto	**c.** $\frac{1}{1,000}$ or .001
4. kiloliters	**d.** 10
5. centi	**e.** grams
6. a car's length	**f.** mm
7. deci	**g.** meters
8. a bathtub's capacity	**h.** $\frac{1}{10}$ or .1
9. millimeter	**i.** $\frac{1}{100}$ or .01
10. a sandwich's weight	**j.** kl
	k. liters
	l. degrees Celsius
	m. 1,000
	n. centigram

ANSWER COLUMN

1. ______
2. ______
3. ______
4. ______
5. ______
6. ______
7. ______
8. ______
9. ______
10. ______

•SECTION 2

DIRECTIONS. Choose the unit that is closest in size to the object named.

1. The length of a pencil is about
a. 20 millimeters **b.** 20 centimeters
c. 20 meters **d.** 20 dekameters

2. A can of soda contains about
a. 35 liters **b.** 35 milliliters
c. 35 centiliters **d.** 35 deciliters

3. The distance a jet plane travels in 1 hour is about
a. 950 centimeters **b.** 9,500 millimeters
c. 9,500 meters **d.** 950 kilometers

1. ______
2. ______
3. ______

Go on to the next page.

4. A soup pot can contain about
   a. 6 liters
   b. 6 hectoliters
   c. 6 dekaliters
   d. 6 centiliters

5. The height of a tall tree can be
   a. 30 meters
   b. 300 millimeters
   c. 30 dekameters
   d. 3 decimeters

ANSWER COLUMN

4. ______

5. ______

•SECTION 3

DIRECTIONS. Choose the unit that is closest in size to the object named.

1. A can of soup weighs about
   a. 30 kilograms
   b. 300 grams
   c. 30 milligrams
   d. 300 centigrams

2. The temperature of a warm bath is about
   a. 95°C
   b. 43°C
   c. 110°C
   d. 15°C

3. The height of a giraffe is about
   a. 487 centimeters
   b. 487 meters
   c. 487 millimeters
   d. 487 dekameters

4. The temperature when skiing is about
   a. 20°C
   b. −10°C
   c. 10°C
   d. −60°C

5. The amount of cheese on a cheeseburger can be
   a. 25 grams
   b. 25 kilograms
   c. 25 hectograms
   d. 25 milligrams

1. ______

2. ______

3. ______

4. ______

5. ______

Give the completed Goal Evaluation to your instructor.

Section 1 ______ %
2 ______ %
3 ______ %
☐ NEXT GOAL
☐ Plan review

← WRITE YOUR SCORE ON YOUR TRACK RECORD

The checked box tells you what to do next.

NAME ....

DATE ....

ALTERNATE GOAL EVALUATION

GOAL 14 **RENAMING SI UNITS**

Solve each of the following problems. Show all work. Write the letter of the correct answer in the answer column at the right.

DIRECTIONS. Choose the number that makes each statement true.

1. __?__ liters = 7,500 milliliters
   a. 7,500,000
   b. 750
   c. 75
   d. 7.5

2. 3.2 kilometers = __?__ meters
   a. 32
   b. 320
   c. 3,200
   d. 32,000

3. __?__ grams = 2.75 dekagrams
   a. .275
   b. 27.5
   c. 275
   d. 2,750

4. 47 hectoliters = __?__ liters
   a. 4,700
   b. 47,000
   c. 4.7
   d. .47

5. 140 deciliters = __?__ liters
   a. 1.4
   b. .14
   c. 14
   d. 1,400

6. 215 centigrams = __?__ grams
   a. 21,500
   b. .215
   c. 2.15
   d. 2,150

7. 87 kilograms = __?__ grams
   a. 8.7
   b. 870
   c. 8,700
   d. 87,000

8. __?__ liters = 612 dekaliters
   a. 61.2
   b. 6,120
   c. 6.12
   d. 61,200

ANSWER COLUMN

1. ______
2. ______
3. ______
4. ______
5. ______
6. ______
7. ______
8. ______

Go on to the next page.

9. ___?___ meters = 1.46 decimeters

a. 14.6
b. 146
c. .0146
d. .146

10. 3,450 milliliters = ___?___ liters

a. 3,450,000
b. 345,000
c. 3.45
d. .345

ANSWER COLUMN

9. ______
10. ______

Give the completed Goal Evaluation to your instructor.

_____% ← WRITE YOUR SCORE ON YOUR TRACK RECORD

☐ NEXT GOAL
☐ Plan review

The checked box tells you what to do next.

NAME ....

DATE ....

ALTERNATE GOAL EVALUATION

GOAL 15 MEASURING WITH U.S. CUSTOMARY AND SI UNITS

Solve each of the following problems. Show all work. Write the letter of the correct answer in the answer column at the right.

DIRECTIONS. Use a standard inch ruler to measure the distance between the marks on each figure to the nearest $\frac{1}{16}$ inch.

Note: Measure from the left side of one dot to the left side of the next dot.

ANSWER COLUMN

1. ______
2. ______
3. ______
4. ______
5. ______

1.

a. $1\frac{1}{8}$ in.
b. $1\frac{3}{16}$ in.
c. $1\frac{1}{4}$ in.
d. $1\frac{3}{8}$ in.

2.

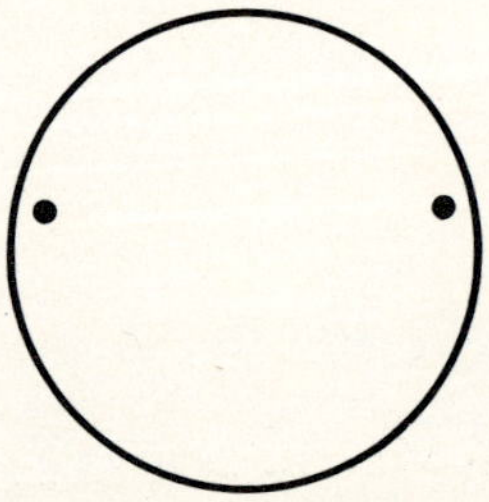

a. $\frac{15}{16}$ in.
b. $1\frac{1}{8}$ in.
c. $1\frac{1}{16}$ in.
d. 1 in.

3.

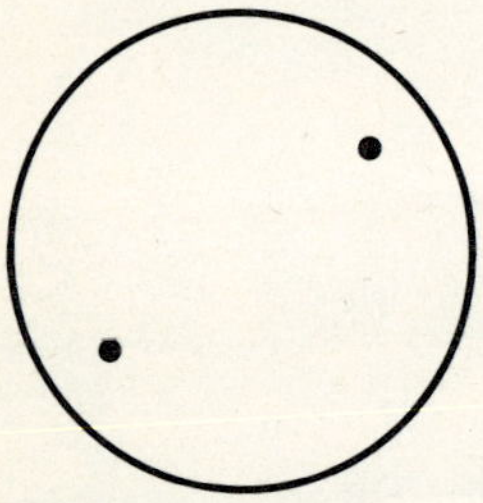

a. $\frac{3}{4}$ in.
b. $\frac{15}{16}$ in.
c. $\frac{13}{16}$ in.
d. $\frac{7}{8}$ in.

4.

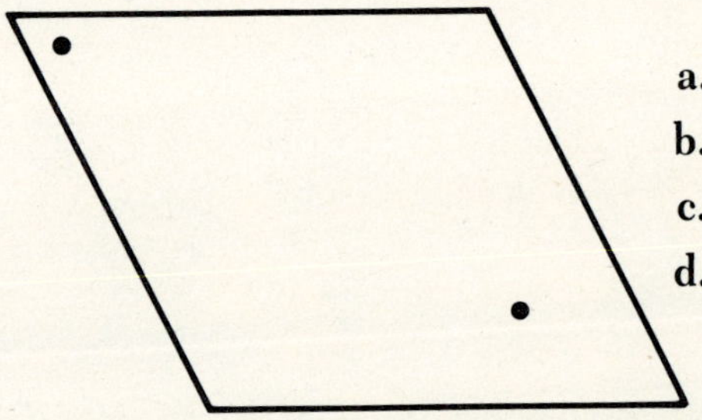

a. $1\frac{15}{16}$ in.
b. $1\frac{13}{16}$ in.
c. $1\frac{3}{4}$ in.
d. $1\frac{7}{8}$ in.

5.

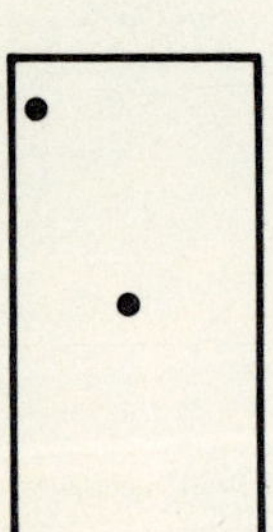

a. $\frac{9}{16}$ in.
b. $\frac{1}{2}$ in.
c. $\frac{7}{16}$ in.
d. $\frac{5}{8}$ in.

Go on to the next page.

ANSWER COLUMN

6. ______
7. ______
8. ______
9. ______
10. ______

DIRECTIONS. Use a standard centimeter ruler to measure the distance between the marks on each figure to the nearest millimeter.

Note: Measure from the left side of one dot to the left side of the next dot.

6.

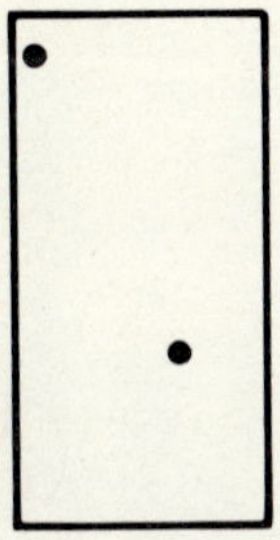

a. 22 mm
b. 23 mm
c. 24 mm
d. 21 mm

7.

a. 12 mm
b. 13 mm
c. 14 mm
d. 11 mm

8.

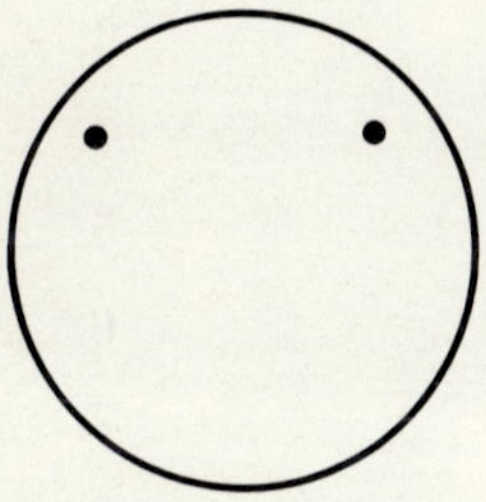

a. 20 mm
b. 19 mm
c. 18 mm
d. 21 mm

9.

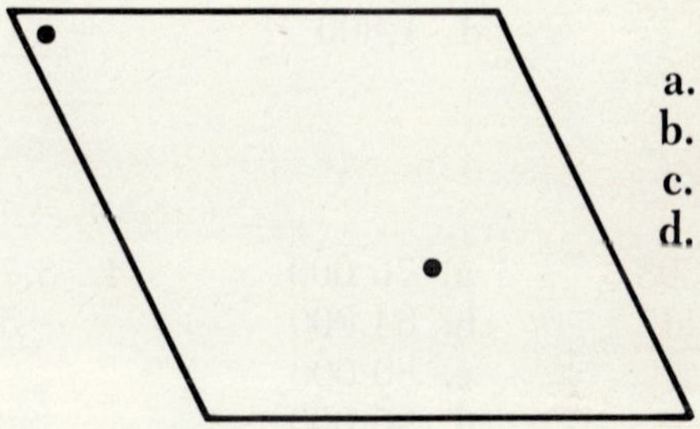

a. 36 mm
b. 37 mm
c. 38 mm
d. 35 mm

10.

a. 28 mm
b. 30 mm
c. 27 mm
d. 29 mm

Give your completed Goal Evaluation to your instructor.

_____% ← WRITE YOUR SCORE ON YOUR TRACK RECORD

☐ NEXT GOAL
☐ Plan review

The checked box tells you what to do next.

NAME ....

DATE ....

ALTERNATE GOAL EVALUATION

GOAL 16 **ESTIMATING ANSWERS**

Solve each of the following problems. Show all work. Write the letter of the correct answer in the answer column at the right.

DIRECTIONS. Estimate the answer to each problem. Don't find the exact answer. Choose the estimate that seems most reasonable.

1.
```
  790
  385
   86
 + 98
```
a. 1,200 b. 1,400 c. 1,600 d. 1,800

2.
```
 $689.17
  231.82
   95.37
 + 17.32
```
a. $1,000 b. $800 c. $1,400 d. $1,200

3.
```
  91,898
 − 8,117
```
a. 70,000 b. 84,000 c. 80,000 d. 90,000

4.
```
 8,796
 − 522
```
a. 7,500 b. 8,500 c. 8,300 d. 8,000

5. During July, Matthew Perez worked on 162 programs. By the end of the month 47 were not completed. How many programs were completed?

a. 100 b. 110
c. 130 d. 140

6.
```
 63,715
  × 310
```
a. 200,000 b. 2,000,000 c. 20,000,000 d. 200,000,000

7.
```
  397
 × 561
```
a. 20,000 b. 200,000 c. 2,000,000 d. 20,000,000

8. 6,472)3,184,962

a. 500,000 b. 50,000 c. 5,000 d. 500

9. 37)227,614

a. 600 b. 6,000 c. 60,000 d. 600,000

ANSWER COLUMN

1. ______
2. ______
3. ______
4. ______
5. ______
6. ______
7. ______
8. ______
9. ______

Go on to the next page.

10. During one week Friend's Office Equipment made 53 sales, totaling $336,487. What was the average amount of each sale?

a. $6,000 b. $60,000
c. $60 d. $600

ANSWER COLUMN

10. ______

Give the completed Goal Evaluation to your instructor.

____%	← WRITE YOUR SCORE ON YOUR TRACK RECORD
☐ NEXT GOAL	
☐ Plan review	

The checked box tells you what to do next.

NAME ....

DATE ....

ALTERNATE GOAL EVALUATION

GOAL 17

READING TABLES AND GRAPHS

Solve each of the following problems. Show all work. Write the letter of the correct answer in the answer column at the right.

•SECTION 1

DIRECTIONS. Use the tables to answer these questions.

1. According to Table 1, what is the tax on $.80?
   a. $.87 **b.** $.14
   c. $.06 **d.** $.07

2. According to Table 1, what is the tax on $1.31?
   a. $.11 **b.** $.03
   c. $.09 **d.** $.10

3. Refer to Table 2 (page 173). What is the tax on weekly pay of $160 if 5 allowances are claimed?
   a. $8.80 **b.** $4.40
   c. $7.30 **d.** $5.80

4. Refer to Table 2 (page 173). What is the tax on weekly pay of $175 if 4 allowances are claimed?
   a. $10.40 **b.** $27.00
   c. $24.90 **d.** $12.20

TABLE 1 8% SALES TAX TABLE

Amount of Sale	Tax
$ .01 to $ .10	$.00
.11 to .17	.01
.18 to .29	.02
.30 to .42	.03
.43 to .54	.04
.55 to .67	.05
.68 to .79	.06
.80 to .92	.07
.93 to 1.06	.08
1.07 to 1.18	.09
1.19 to 1.31	.10
1.32 to 1.43	.11
1.44 to 1.56	.12
1.57 to 1.68	.13
1.69 to 1.81	.14
1.82 to 1.93	.15
1.94 to 2.06	.16
2.07 to 2.18	.17
2.19 to 2.31	.18
2.32 to 2.43	.19

ANSWER COLUMN

1. ______
2. ______
3. ______
4. ______
5. ______
6. ______

•SECTION 2

DIRECTIONS. Use the graphs on pages 173 and 174 to answer these questions.

5. Refer to the graph of Merlin's Home Computers. How many microcomputers were sold in April?
   a. 19,000 **b.** 19,750
   c. 22,000 **d.** 20,000

6. Refer to the graph of Merlin's Home Computers. In which month were sales 22,000?
   a. July **b.** May
   c. August **d.** September

Go on to the next page.

TABLE 2

SINGLE Persons — WEEKLY Payroll Period

And the wages are—		And the number of withholding allowances claimed is—										
At least	But less than	0	1	2	3	4	5	6	7	8	9	10
		The amount of income tax to be withheld shall be—										
$135	$140	$17.80	$14.10	$10.50	$7.40	$4.70	$2.00	$0	$0	$0	$0	$0
140	145	18.70	15.10	11.40	8.20	5.40	2.70	0	0	0	0	0
145	150	19.70	16.00	12.40	9.00	6.10	3.40	.70	0	0	0	0
150	160	21.10	17.50	13.80	10.20	7.20	4.40	1.70	0	0	0	0
160	170	23.00	19.40	15.70	12.10	8.80	5.80	3.10	.40	0	0	0
170	180	24.90	21.30	17.60	14.00	10.40	7.30	4.50	1.80	0	0	0
180	190	27.00	23.20	19.50	15.90	12.20	8.90	5.90	3.20	.50	0	0
190	200	29.40	25.10	21.40	17.80	14.10	10.50	7.40	4.60	1.90	0	0
200	210	31.80	27.20	23.30	19.70	16.00	12.40	9.00	6.00	3.30	.60	0
210	220	34.20	29.60	25.20	21.60	17.90	14.30	10.60	7.50	4.70	2.00	0
220	230	36.60	32.00	27.40	23.50	19.80	16.20	12.50	9.10	6.10	3.40	.80
230	240	39.00	34.40	29.80	25.40	21.70	18.10	14.40	10.70	7.70	4.80	2.20
240	250	41.40	36.80	32.20	27.60	23.60	20.00	16.30	12.60	9.30	6.20	3.60
250	260	43.80	39.20	34.60	30.00	25.50	21.90	18.20	14.50	10.90	7.80	5.00
260	270	46.20	41.60	37.00	32.40	27.80	23.80	20.10	16.40	12.80	9.40	6.40
270	280	48.90	44.00	39.40	34.80	30.20	25.70	22.00	18.30	14.70	11.00	7.90
280	290	51.80	46.40	41.80	37.20	32.60	28.00	23.90	20.20	16.60	12.90	9.50
290	300	54.70	49.10	44.20	39.60	35.00	30.40	25.80	22.10	18.50	14.80	11.20
300	310	57.60	52.00	46.60	42.00	37.40	32.80	28.10	24.00	20.40	16.70	13.10
310	320	60.50	54.90	49.40	44.40	39.80	35.20	30.50	25.90	22.30	18.60	15.00
320	330	63.40	57.80	52.30	46.80	42.20	37.60	32.90	28.30	24.20	20.50	16.90
330	340	66.50	60.70	55.20	49.60	44.60	40.00	35.30	30.70	26.10	22.40	18.80
340	350	69.70	63.60	58.10	52.50	47.00	42.40	37.70	33.10	28.50	24.30	20.70
350	360	72.90	66.70	61.00	55.40	49.80	44.80	40.10	35.50	30.90	26.30	22.60
360	370	76.10	69.90	63.90	58.30	52.70	47.20	42.50	37.90	33.30	28.70	24.50
370	380	79.30	73.10	66.90	61.20	55.60	50.00	44.90	40.30	35.70	31.10	26.50
380	390	82.50	76.30	70.10	64.10	58.50	52.90	47.40	42.70	38.10	33.50	28.90
390	400	85.70	79.50	73.30	67.20	61.40	55.80	50.30	45.10	40.50	35.90	31.30
400	410	88.90	82.70	76.50	70.40	64.30	58.70	53.20	47.60	42.90	38.30	33.70
410	420	92.10	85.90	79.70	73.60	67.40	61.60	56.10	50.50	45.30	40.70	36.10

Merlin's Home Computers
Monthly Sales of Micros
19—

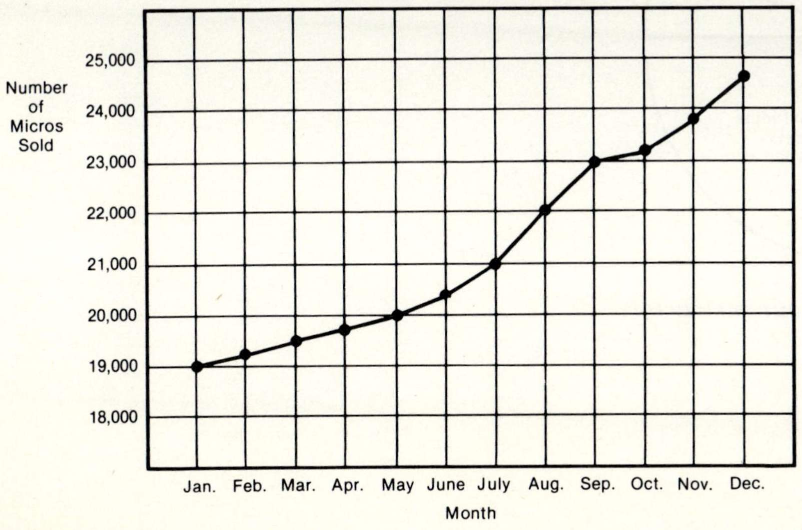

Go on to the next page.

7. Refer to the graph of retail stores. What were the sales in dollars at eating and drinking places?
a. 80 billion **b.** 90 billion
c. 85 billion **d.** 100 billion

8. Refer to the graph of retail stores. How much lower were sales at gasoline service stations than were sales at automotive dealers?
a. 145 billion **b.** 90 billion
c. 55 billion **d.** 100 billion

ANSWER COLUMN

7. ______
8. ______
9. ______
10. ______

Retail Stores Sales by Kind of Business
Period Ending 19—

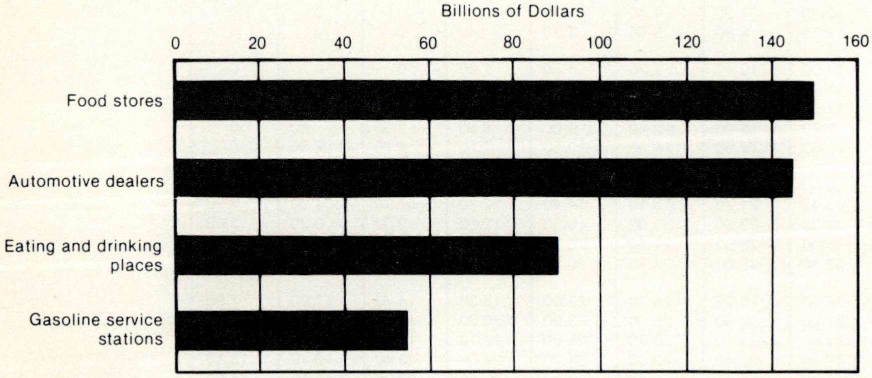

9. Refer to the sales graph of ABC Electronics Company. How many percent points higher were sales of black and white TV sets than were sales of stereos?
a. 15% **b.** 50%
c. 35% **d.** 20%

10. Refer to the sales graph of ABC Electronics Company. What percent of total sales were sales of color and black and white TV sets?
a. 50% **b.** 55%
c. 80% **d.** 35%

Sales of ABC Electronics Company
Year Ending 19—

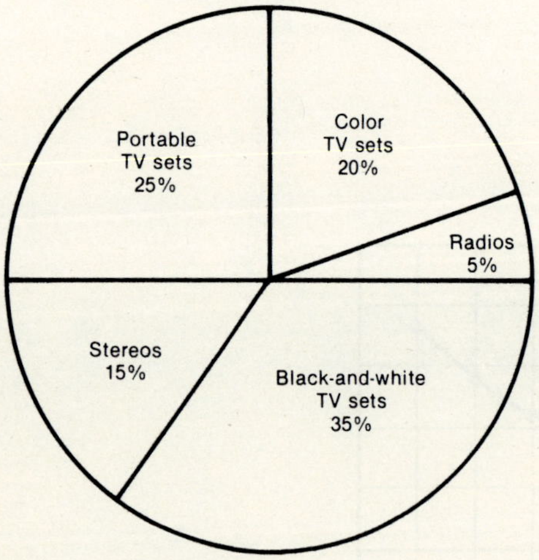

Give the completed Goal Evaluation to your instructor.

Section 1_____%
2_____%
☐ NEXT GOAL
☐ Plan review

WRITE YOUR SCORE ON YOUR TRACK RECORD

The checked box tells you what to do next.

NAME ....

DATE ....

ALTERNATE GOAL EVALUATION

GOAL 18 **USING EQUATIONS AND BUSINESS FORMULAS**

Solve each of the following problems. Show all work. Write the letter of the correct answer in the answer column at the right.

ANSWER COLUMN

1. ______
2. ______
3. ______
4. ______
5. ______

DIRECTIONS. Solve.

1. Masie Lopez borrowed $4,500 to complete her education. She paid $450 interest for a 1-year period. What was the interest rate?

a. 9.9% **b.** 9.1%
c. 10% **d.** 11%

2. Super Sales Outlet marks up paper products $66\frac{2}{3}$% on the cost price. If paper towels cost $.48, what is the markup?

a. $.32 **b.** $.80
c. $.16 **d.** $.16

3. Muriel Seeburg sold a $700 color TV. If she gets a 6% commission on all sales, how much commission did she make?

a. $742 **b.** $42
c. $70 **d.** $658

4. Janet Henderson receives a 5% commission on sales. Her commission was $275. What was the amount of her sales?

a. $5,225 **b.** $55
c. $1,375 **d.** $5,500

5. Cobatop Paper Products sales increased from $22,000 a week to $25,300 a week. What was the percent of increase in sales?

a. 13% **b.** 87%
c. 15% **d.** 115%

Go on to the next page.

ANSWER COLUMN

6. ______
7. ______
8. ______
9. ______
10. ______

6. The sales tax on a $7,250 word processor was $580. What was the tax rate?

a. 108% b. 92%
c. 12.5% d. 8%

7. Since last year Marchison Office Products increased its number of outlets by 25%. If there were 360 outlets last year, how many more outlets are there this year?

a. 450 b. 270
c. 80 d. 90

8. Baker Office Machine Repair Service increased the number of machines repaired each day from 6 to 9. What is the percent of increase?

a. $33\frac{1}{3}$% b. 150%
c. $66\frac{2}{3}$% d. 50%

9. Henry Tudor invested $18,000 for 1 year at 15% interest. How much interest did he get?

a. $1,200 b. $2,700
c. $270 d. $20,700

10. Industrial Steel Products bought $4,200 worth of new production equipment. A 7% discount was offered for payment made in cash. How much is the discount?

a. $294 b. $4,494
c. $3,906 d. $600

Give the completed Goal Evaluation to your instructor.

_____% ← WRITE YOUR SCORE ON YOUR TRACK RECORD

☐ NEXT GOAL
☐ Plan review

The checked box tells you what to do next.

NAME ....

DATE ....

ALTERNATE GOAL EVALUATION

GOAL 19

SOLVING WORD PROBLEMS

Solve each of the following problems. Show all work. Write the letter of the correct answer in the answer column at the right.

DIRECTIONS. If there are not enough facts to answer a problem, choose the answer "more facts needed."

ANSWER COLUMN

1. ______

2. ______

3. ______

1. The office budget contains $560 for servicing office machines. If service calls average $35 each, how many service calls will this cover?
a. more facts needed **b.** 18
c. 16 **d.** 14

2. Simpson Office Furniture sells filing equipment at their cost plus a $37\frac{1}{2}\%$ markup. What is the markup on a file cabinet that cost them $88?
a. more facts needed **b.** $33
c. $121 **d.** $55

3. Garcia Catering Service makes $.73 profit on each meal sold. In 4 months they sold 27,000 meals. What was their profit?
a. more facts needed **b.** $6,750
c. $4,927.50 **d.** $19,710

Go on to the next page.

4. Valley Savings Bank charges $.25 per check to customers who do not have a savings account with Valley. John Agagianian wrote 10 checks for $1,528. What was he charged?
   a. more facts needed b. $1,530.50
   c. $2.50 d. $0

ANSWER COLUMN

4. ______
5. ______
6. ______
7. ______

5. Last year Charles Martel made $12,000 in salary, $15,000 in commission, and $3,000 in profit sharing. This year his salary is $15,000. What is the percent of increase in salary?
   a. more facts needed b. 80%
   c. 25% d. 10%

6. Nicholas Romanoff gets $.17 for each tuner he assembles. In 1 week he assembled 189 tuners. He worked 5 days that week. What were his weekly earnings?
   a. more facts needed b. $160.65
   c. $32.13 d. $945

7. Henry Tudor worked 17 hours overtime in 1 week. If he gets $1\frac{1}{2}$ times his regular pay for overtime, how many hours' pay did he get that week?
   a. more facts needed b. 57 hours
   c. 25.5 hours d. 65.5 hours

Go on to the next page.

ANSWER COLUMN

8. ______

9. ______

10. ______

8. Elizabeth Windsor's profits are $37\frac{1}{2}\%$ of the sales prices of her company's products. Last month sales were $8,000. What was her profit?

a. more facts needed **b.** $3,000
c. $5,000 **d.** $11,000

9. Charles Stuart earned $24,000 last year. He had to pay 30% in taxes. What did he actually receive?

a. more facts needed **b.** $31,200
c. $7,200 **d.** $16,800

10. Ivan Grozny borrowed $2,000 for 1 year at 15% interest. How much did he owe when the loan was due?

a. more facts needed **b.** $1,700
c. $300 **d.** $2,300

Give the completed Goal Evaluation to your instructor.

____% ← WRITE YOUR SCORE ON YOUR TRACK RECORD
☐ NEXT GOAL
☐ Plan review

The checked box tells you what to do next.

NAME ....

DATE ....

ALTERNATE GOAL EVALUATION

GOAL 20

RENAMING COMMON FRACTIONS TO SIMPLEST (LOWEST) TERMS

Solve each of the following problems. Show all work. Write the letter of the correct answer in the answer column at the right.

DIRECTIONS. Rename each fraction to lowest terms.

1. $\frac{81}{90}$

- **a.** $\frac{8}{9}$
- **b.** $\frac{27}{30}$
- **c.** $\frac{9}{10}$
- **d.** $\frac{2}{3}$

2. $\frac{12}{40}$

- **a.** $\frac{6}{20}$
- **b.** $\frac{3}{10}$
- **c.** $\frac{1}{4}$
- **d.** $\frac{3}{20}$

3. $\frac{2,100}{5,600}$

- **a.** $\frac{21}{56}$
- **b.** $\frac{3}{8}$
- **c.** $\frac{3}{7}$
- **d.** $\frac{2}{5}$

4. $\frac{16}{52}$

- **a.** $\frac{4}{13}$
- **b.** $\frac{1}{3}$
- **c.** $\frac{1}{5}$
- **d.** $\frac{8}{26}$

5. $\frac{84}{96}$

- **a.** 21.24
- **b.** $\frac{28}{32}$
- **c.** $\frac{8}{9}$
- **d.** $\frac{7}{8}$

ANSWER COLUMN

1. ______
2. ______
3. ______
4. ______
5. ______

Go on to the next page.

6. $\frac{300}{750}$

a. $\frac{2}{5}$

b. $\frac{3}{7}$

c. $\frac{10}{25}$

d. $\frac{1}{3}$

7. $\frac{114}{192}$

a. $\frac{11}{19}$

b. $\frac{57}{96}$

c. $\frac{19}{32}$

d. $\frac{38}{64}$

ANSWER COLUMN

6. ______

7. ______

8. ______

9. ______

10. ______

8. $\frac{5{,}200}{7{,}800}$

a. $\frac{26}{39}$

b. $\frac{5}{7}$

c. $\frac{52}{78}$

d. $\frac{2}{3}$

9. $\frac{18}{108}$

a. $\frac{6}{36}$

b. $\frac{1}{10}$

c. $\frac{9}{54}$

d. $\frac{1}{6}$

10. $\frac{660}{1{,}560}$

a. $\frac{66}{156}$

b. $\frac{11}{26}$

c. $\frac{22}{32}$

d. $\frac{2}{5}$

Give the completed Goal Evaluation to your instructor.

_____% ←	WRITE YOUR SCORE ON YOUR TRACK RECORD
☐ NEXT GOAL	
☐ Plan review	

The checked box tells you what to do next.

NAME ....

DATE ....

ALTERNATE GOAL EVALUATION

GOAL 21 **RENAMING MIXED NUMBERS AS COMMON FRACTIONS**

Solve each of the following problems. Show all work. Write the letter of the correct answer in the answer column at the right.

DIRECTIONS. Choose the number that makes each statement true.

1. $23 = \underline{\ ?\ }$

a. $\frac{115}{5}$
b. $4\frac{3}{5}$
c. $\frac{19}{5}$
d. $\frac{24}{5}$

2. $8\frac{1}{3} = \underline{\ ?\ }$

a. $8\frac{1}{3}$
b. $\frac{9}{3}$
c. $\frac{25}{3}$
d. $\frac{8}{3}$

3. $6\frac{5}{6} = \underline{\ ?\ }$

a. $\frac{65}{6}$
b. $\frac{41}{6}$
c. $\frac{11}{6}$
d. $\frac{30}{6}$

4. $11\frac{3}{8} = \underline{\ ?\ }$

a. $\frac{113}{8}$
b. $\frac{91}{8}$
c. $\frac{14}{8}$
d. $\frac{33}{8}$

5. $27\frac{4}{7} = \underline{\ ?\ }$

a. $\frac{193}{7}$
b. $\frac{274}{7}$
c. $\frac{31}{7}$
d. $\frac{108}{7}$

ANSWER COLUMN

1. ______
2. ______
3. ______
4. ______
5. ______

Go on to the next page.

DIRECTIONS. Rename these improper fractions as mixed numbers.

ANSWER COLUMN

6. ______
7. ______
8. ______
9. ______
10. ______

6. $\frac{511}{8}$ = ?
a. $51\frac{1}{8}$
b. $62\frac{9}{8}$
c. $63\frac{7}{8}$
d. $65\frac{1}{8}$

7. $\frac{174}{7}$ = ?
a. $17\frac{4}{7}$
b. $24\frac{6}{7}$
c. $25\frac{4}{7}$
d. $20\frac{4}{7}$

8. $\frac{193}{5}$ = ?
a. $19\frac{3}{5}$
b. $38\frac{3}{5}$
c. $36\frac{3}{5}$
d. $39\frac{1}{5}$

9. $\frac{251}{6}$ = ?
a. $25\frac{1}{6}$
b. $48\frac{1}{2}$
c. $46\frac{1}{2}$
d. $41\frac{5}{6}$

10. $\frac{327}{9}$ = ?
a. $36\frac{1}{3}$
b. $37\frac{1}{9}$
c. $37\frac{1}{3}$
d. $32\frac{7}{9}$

Give the completed Goal Evaluation to your instructor.

____% ← WRITE YOUR SCORE ON YOUR TRACK RECORD
☐ NEXT GOAL
☐ Plan review

The checked box tells you what to do next.

NAME ....

DATE ....

ALTERNATE GOAL EVALUATION

GOAL 22 MULTIPLYING COMMON FRACTIONS

Solve each of the following problems. Show all work. Write the letter of the correct answer in the answer column at the right.

DIRECTIONS. Multiply.

1. $\$75 \times 1\frac{3}{5}$

a. $\$75\frac{3}{5}$
b. $120
c. $\$48\frac{1}{8}$
d. $45

2. $\frac{5}{8} \times \frac{2}{5}$

a. $\frac{10}{40}$ or $\frac{1}{4}$
b. $1\frac{9}{16}$
c. $\frac{7}{13}$
d. $\frac{16}{25}$

3. $49 \times 1\frac{2}{5}$

a. 35
b. $49\frac{2}{5}$
c. $19\frac{3}{5}$
d. $68\frac{3}{5}$

4. $\frac{2}{7} \times \frac{2}{3}$

a. $\frac{2}{5}$
b. $2\frac{1}{3}$
c. $\frac{3}{7}$
d. $\frac{4}{21}$

5. $\frac{3}{10} \times \frac{4}{5}$

a. $\frac{7}{15}$
b. $\frac{12}{50}$ or $\frac{6}{25}$
c. $\frac{3}{8}$
d. $1\frac{1}{3}$

ANSWER COLUMN

1. ______
2. ______
3. ______
4. ______
5. ______

Give the completed Goal Evaluation to your instructor.

_____% ← ☐ NEXT GOAL ☐ Plan review	WRITE YOUR SCORE ON YOUR TRACK RECORD

The checked box tells you what to do next.

NAME ....

DATE ....

ALTERNATE GOAL EVALUATION

GOAL 23

DIVIDING WITH COMMON FRACTIONS

Solve each of the following problems. Show all work. Write the letter of the correct answer in the answer column at the right.

DIRECTIONS. Divide.

1. $3{,}807 \div 1\frac{1}{8}$

a. 3,384
b. $4{,}282\frac{7}{8}$
c. $475\frac{7}{8}$
d. $52\frac{7}{8}$

2. $9{,}921 \div \frac{3}{4}$

a. $7{,}741\frac{3}{4}$
b. $826\frac{3}{4}$
c. 13,228
d. 3,307

3. $1{,}851 \div 2\frac{1}{3}$

a. 4,326
b. $927\frac{1}{3}$
c. $794\frac{4}{7}$
d. $88\frac{2}{7}$

4. $9{,}218 \div 11\frac{1}{2}$

a. $801\frac{13}{23}$
b. 419
c. 50,699
d. $838\frac{1}{2}$

5. $2{,}543 \div \frac{7}{8}$

a. $324\frac{7}{8}$
b. $2{,}906\frac{2}{7}$
c. $2{,}225\frac{1}{8}$
d. $371\frac{2}{7}$

ANSWER COLUMN

1. ______
2. ______
3. ______
4. ______
5. ______

Go on to the next page.

ANSWER COLUMN

6. ______
7. ______
8. ______
9. ______
10. ______

6. $3{,}108 \div 7\frac{1}{3}$

a. $47\frac{1}{11}$

b. 22,792

c. $423\frac{9}{11}$

d. $444\frac{9}{11}$

7. $3{,}618 \div 2\frac{2}{5}$

a. $60\frac{3}{10}$

b. $8{,}683\frac{2}{5}$

c. $1{,}809\frac{2}{5}$

d. $1{,}507\frac{1}{2}$

8. $4{,}361 \div \frac{2}{7}$

a. $311\frac{1}{2}$

b. $6\frac{2}{3}$

c. 1,246

d. $15{,}263\frac{1}{2}$

9. $64.35 ÷ $1\frac{4}{5}$

a. $115.65

b. $35.75

c. $1.43

d. $51.48

10. $96.30 ÷ $\frac{5}{6}$

a. $115.56

b. $3.21

c. $80.25

d. $16.05

Give the completed Goal Evaluation to your instructor.

____% ← WRITE YOUR SCORE ON YOUR TRACK RECORD
☐ NEXT GOAL
☐ Plan review

The checked box tells you what to do next.

NAME ....

DATE ....

ALTERNATE GOAL EVALUATION

GOAL 24 **ADDING COMMON FRACTIONS**

Solve each of the following problems. Show all work. Write the letter of the correct answer in the answer column at the right.

DIRECTIONS. Add.

ANSWER COLUMN

1. ______
2. ______
3. ______
4. ______
5. ______

1. $\frac{11}{5} + \frac{2}{5} + \frac{4}{5} + \frac{7}{5}$

a. $\frac{24}{25}$ b. $2\frac{2}{5}$ c. $4\frac{4}{5}$ d. $1\frac{1}{5}$

2. $\frac{4}{3} + \frac{7}{3} + \frac{1}{3} + \frac{5}{3}$

a. $5\frac{2}{3}$ b. $1\frac{5}{12}$ c. $2\frac{5}{6}$ d. $1\frac{8}{9}$

3. $\frac{5}{6} + \frac{11}{12} + \frac{9}{6} + \frac{7}{12}$

a. $\frac{8}{9}$ b. $2\frac{2}{3}$ c. $1\frac{7}{9}$ d. $3\frac{5}{6}$

4. $\frac{1}{3} + \frac{3}{2} + \frac{2}{4} + \frac{5}{6}$

a. $3\frac{1}{6}$ b. $\frac{11}{15}$ c. $3\frac{1}{4}$ d. $\frac{11}{12}$

5. $\frac{5}{8} + \frac{4}{3} + \frac{5}{6} + \frac{1}{2}$

a. $\frac{15}{19}$ b. $3\frac{1}{3}$ c. $3\frac{7}{24}$ d. $\frac{2}{3}$

Give the completed Goal Evaluation to your instructor.

____% ← WRITE YOUR SCORE ON YOUR TRACK RECORD
☐ NEXT GOAL
☐ Plan review

The checked box tells you what to do next.

NAME ....

DATE ....

ALTERNATE GOAL EVALUATION

GOAL 25

SUBTRACTING COMMON FRACTIONS

Solve each of the following problems. Show all work. Write the letter of the correct answer in the answer column at the right.

DIRECTIONS. Subtract.

1. $\frac{5}{10} - \frac{2}{5}$

a. $\frac{1}{10}$
b. $\frac{3}{10}$
c. $\frac{3}{50}$
d. $\frac{3}{5}$

2. $\frac{8}{9} - \frac{4}{6}$

a. $\frac{4}{9}$
b. $\frac{2}{9}$
c. $\frac{4}{3}$
d. $\frac{12}{54}$ or $\frac{2}{9}$

3. $\frac{7}{12} - \frac{1}{3}$

a. $\frac{3}{12}$ or $\frac{1}{4}$
b. $\frac{1}{2}$
c. $\frac{7}{9}$
d. $\frac{1}{6}$

4. $\frac{9}{14} - \frac{3}{7}$

a. $\frac{6}{7}$
b. $\frac{6}{14}$ or $\frac{3}{7}$
c. $\frac{6}{98}$ or $\frac{3}{49}$
d. $\frac{3}{14}$

5. $\frac{5}{8} - \frac{3}{5}$

a. $\frac{2}{3}$
b. $\frac{2}{40}$ or $\frac{1}{20}$
c. $\frac{1}{13}$
d. $\frac{1}{40}$

ANSWER COLUMN

1. ______
2. ______
3. ______
4. ______
5. ______

Go on to the next page.

6. $6\frac{5}{9} - 3\frac{3}{6}$

a. $3\frac{2}{3}$

b. $3\frac{1}{18}$

c. $2\frac{1}{18}$

d. $3\frac{2}{18}$ or $3\frac{1}{9}$

7. $7\frac{5}{12} - 3\frac{2}{3}$

a. $4\frac{3}{9}$

b. $4\frac{9}{12}$ or $4\frac{3}{4}$

c. $4\frac{3}{12}$ or $4\frac{1}{4}$

d. $3\frac{9}{12}$ or $3\frac{3}{4}$

ANSWER COLUMN

6. ______

7. ______

8. ______

9. ______

10. ______

8. $7\frac{5}{8} - 5\frac{3}{4}$

a. $2\frac{2}{4}$ or $2\frac{1}{2}$

b. $1\frac{7}{8}$

c. $2\frac{2}{8}$ or $2\frac{1}{4}$

d. $2\frac{7}{8}$

9. $10\frac{7}{10} - 8\frac{4}{5}$

a. $2\frac{3}{5}$

b. $2\frac{3}{10}$

c. $1\frac{9}{10}$

d. $2\frac{9}{10}$

10. $12\frac{3}{10} - 8\frac{1}{2}$

a. $3\frac{8}{10}$ or $3\frac{4}{5}$

b. $4\frac{2}{8}$ or $4\frac{1}{4}$

c. $4\frac{2}{10}$ or $4\frac{1}{5}$

d. $4\frac{8}{10}$ or $4\frac{4}{5}$

Give the completed Goal Evaluation to your instructor.

______% ← WRITE YOUR SCORE ON YOUR TRACK RECORD
☐ NEXT GOAL
☐ Plan review

The checked box tells you what to do next.

TRACK RECORD of ______________________________

Where To Find It →	SKILLS TEST	LEARNING GUIDE	TEXTBOOK	INSTRUCTOR'S MANUAL AND KEY	SKILLS TEST
	BEGINNING SCORE	PRECHECK	WORK TO DO	GOAL EVALUATION	FINAL SCORE
GOAL 1	Reading and Writing Numbers ______% ☐ Take Precheck on *Learning Guide* page 1	______% ☐ NEXT GOAL	☐ Text page 1	______% ☐ NEXT GOAL ☐ Plan review	______%
GOAL 2	Rounding Off Numbers ______% ☐ Take Precheck on *Learning Guide* page 5	______% ☐ NEXT GOAL	☐ Text page 25	______% ☐ NEXT GOAL ☐ Plan review	______%
GOAL 3	Adding Whole Numbers and Money Amounts ______% ☐ Take Precheck on *Learning Guide* page 9	Section 1______% 2______% ☐ NEXT GOAL	☐ Text page 39 ☐ Text page 53	Section 1______% 2______% ☐ NEXT GOAL ☐ Plan review	______%
GOAL 4	Subtracting Whole Numbers ______% ☐ Take Precheck on *Learning Guide* page 15	Section 1______% 2______% ☐ NEXT GOAL	☐ Text page 68 ☐ Text page 76	Section 1______% 2______% ☐ NEXT GOAL ☐ Plan review	______%
GOAL 5	Multiplying Whole Numbers ______% ☐ Take Precheck on *Learning Guide* page 25	Section 1______% 2______% ☐ NEXT GOAL	☐ Text page 88 ☐ Text page 98	Section 1______% 2______% ☐ NEXT GOAL ☐ Plan review	______%
GOAL 6	Dividing Whole Numbers ______% ☐ Take Precheck on *Learning Guide* page 31	Section 1______% 2______% ☐ NEXT GOAL	☐ Text page 114 ☐ Text page 133	Section 1______% 2______% ☐ NEXT GOAL ☐ Plan review	______%
GOAL 7	Placing the Decimal Point ______% ☐ Take Precheck on *Learning Guide* page 37	Section 1______% 2______% ☐ NEXT GOAL	☐ Text page 149 ☐ Text page 161	Section 1______% 2______% ☐ NEXT GOAL ☐ Plan review	______%
GOAL 8	Writing Common Fractions ______% ☐ Take Precheck on *Learning Guide* page 43	______% ☐ NEXT GOAL	☐ Text page 173	______% ☐ NEXT GOAL ☐ Plan review	______%
GOAL 9	Renaming Common Fractions ______% ☐ Take Precheck on *Learning Guide* page 53	______% ☐ NEXT GOAL	☐ Text page 189	______% ☐ NEXT GOAL ☐ Plan review	______%
GOAL 10	Working With Time ______% ☐ Take Precheck on *Learning Guide* page 57	Section 1______% 2______% ☐ NEXT GOAL	☐ Text page 205 ☐ Text page 221	Section 1______% 2______% ☐ NEXT GOAL ☐ Plan review	______%
GOAL 11	Identifying U.S. Customary Units ______% ☐ Take Precheck on *Learning Guide* page 63	______% ☐ NEXT GOAL	☐ Text page 233	______% ☐ NEXT GOAL ☐ Plan review	______%
GOAL 12	Renaming U.S. Customary Units ______% ☐ Take Precheck on *Learning Guide* page 67	______% ☐ NEXT GOAL	☐ Text page 252	______% ☐ NEXT GOAL ☐ Plan review	______%

TRACK RECORD of ______________________________

Where To Find It →	SKILLS TEST	LEARNING GUIDE	TEXTBOOK	INSTRUCTOR'S MANUAL AND KEY	SKILLS TEST
	BEGINNING SCORE	PRECHECK	WORK TO DO	GOAL EVALUATION	FINAL SCORE
GOAL 13	Identifying SI Units ______% ☐ Take Precheck on *Learning Guide* page 83	Section 1______% 2______% 3______% ☐ NEXT GOAL	☐ Text page 269 ☐ Text page 281 ☐ Text page 293	Section 1______% 2______% 3______% ☐ NEXT GOAL ☐ Plan review	______%
GOAL 14	Renaming SI Units ______% ☐ Take Precheck on *Learning Guide* page 91	______% ☐ NEXT GOAL	☐ Text page 302	______% ☐ NEXT GOAL ☐ Plan review	______%
GOAL 15	Measuring With Rulers ______% ☐ Take Precheck on *Learning Guide* page 95	______% ☐ NEXT GOAL	☐ Text page 314	______% ☐ NEXT GOAL ☐ Plan review	______%
GOAL 16	Estimating Answers ______% ☐ Take Precheck on *Learning Guide* page 99	______% ☐ NEXT GOAL	☐ Text page 336	______% ☐ NEXT GOAL ☐ Plan review	______%
GOAL 17	Reading Table and Graphs ______% ☐ Take Precheck on *Learning Guide* page 103	Section 1______% 2______% ☐ NEXT GOAL	☐ Text page 352 ☐ Text page 374	Section 1______% 2______% ☐ NEXT GOAL ☐ Plan review	______%
GOAL 18	Using Equations ______% ☐ Take Precheck on *Learning Guide* page 117	______% ☐ NEXT GOAL	☐ Text page 387	______% ☐ NEXT GOAL ☐ Plan review	______%
GOAL 19	Solving Word Problems ______% ☐ Take Precheck on *Learning Guide* page 121	______% ☐ NEXT GOAL	☐ Text page 416	______% ☐ NEXT GOAL ☐ Plan review	______%
GOAL 20	Renaming Common Fractions ______% ☐ Take Precheck on *Learning Guide* page 131	______% ☐ NEXT GOAL	☐ Text page 437	______% ☐ NEXT GOAL ☐ Plan review	______%
GOAL 21	Renaming Mixed Numbers ______% ☐ Take Precheck on *Learning Guide* page 135	______% ☐ NEXT GOAL	☐ Text page 456	______% ☐ NEXT GOAL ☐ Plan review	______%
GOAL 22	Multiplying Fractions ______% ☐ Take Precheck on *Learning Guide* page 139	______% ☐ NEXT GOAL	☐ Text page 468	______% ☐ NEXT GOAL ☐ Plan review	______%
GOAL 23	Dividing With Fractions ______% ☐ Take Precheck on *Learning Guide* page 143	______% ☐ NEXT GOAL	☐ Text page 486	______% ☐ NEXT GOAL ☐ Plan review	______%
GOAL 24	Adding Fractions ______% ☐ Take Precheck on *Learning Guide* page 147	______% ☐ NEXT GOAL	☐ Text page 500	______% ☐ NEXT GOAL ☐ Plan review	______%
GOAL 25	Subtracting Fractions ______% ☐ Take Precheck on *Learning Guide* page 157	______% ☐ NEXT GOAL	☐ Text page 519	______% ☐ NEXT GOAL ☐ Plan review	______%

Notes

Notes

Notes

Notes